E. Fumagalli

Statical and Geomechanical Models

Springer-Verlag

New York Wien

Prof. Dr. Ing. EMANUELE FUMAGALLI
Istituto Sperimentale Modelli e Strutture (ISMES)
Bergamo, Italy

© 1973 by Springer-Verlag/Wien
Library of Congress Catalog Card Number 72-83985
Printed in Austria

With 145 Figures

ISBN 0-387-81096-X Springer-Verlag New York-Wien
ISBN 3-211-81096-X Springer-Verlag Wien-New York

To my wife

Acknowledgments

The author wishes to thank the collaborators:

Dr. Ing. Pier Paolo Rossi
Dr. Gianfranco Camponuovo
Mr. Giuseppe Confalonieri

for their valuable support in preparing and arranging the manuscript, and Dr. K. G. Stagg of the University of Wales, Swansea, for the faithful translation of the text.

Foreword

It gives me much pleasure to introduce this work of Prof. Dr. Ing. Fumagalli, a work that covers both the development and present state of the art of structural model techniques.

In my view, in this context, only a lack of understanding of the possibilities offered by experiments on models and sometimes an unjustified suspicion of them have up to now restricted the development that these methods deserve.

I think, in particular, and the many examples quoted in the text bear witness to this, that models today constitute an efficient means of research that have been refined through advances in the methods of reproduction, testing and measurement.

They represent a reliable and above all safe method of investigation, suitable for use in the elastic range and beyond up to failure, as much for historic ancient monuments as for modern works and structures of particularly bold design that are frequently highly redundant.

They are a particularly valuable tool in areas where analytical methods are inadequate, and yet always useful for comparison with analytical results.

Guido Oberti

Preface

I have embarked on writing a text on the techniques of structural models for two basic reasons:

Firstly because I wish to attempt in some measure a personal appreciation of the subject based on more than twenty years experience, insofar as this can be achieved in a logically coherent and complete treatise.

Secondly because I want to define the basic principles established from consolidated experience and set these out in a text suitable for young researchers who intend to devote themselves to this particular discipline.

In this context I would like to point out that the art of construction is not an exact science as normally defined by precise laws and rules.

As an art, although thriving on many branches of science, it is firstly and above all based on the fruit of experience. The styles that aesthetically express, through the more notable structural monuments, the evolution of architectural forms over the centuries and the structural systems that characterise the different epochs through the continued development of structural statics and the better utilisation of constructional materials, clearly demonstrate this conclusion.

As an artist one works within margins of indeterminacy, presumptuously defined as "factors of safety", but which in reality are factors of ignorance of the complex physico-mechanical phenomena that affect construction.

At the present time a complete and reliable evaluation of the safety of a design can only be effected by subjecting the structure to a severe trial that tests its strength up to failure. For mass produced structural elements such tests can be carried out on the prototype; but for single structures, of large dimensions, only tests on a model can provide the basic information on the real factor of safety of the design.

Analytical investigations, normally restricted to elastic behaviour, that satisfactorily fulfil the requirements of the designer when dealing with structures that have been adequately tested by traditional methods are frequently inadequate for bold structures, of new conception, because of the difficulty of formulating a reliable mode of structural behaviour.

Experience has shown me by example that, when a given structure reaches limits of size or slenderness never previously attained, the statical investigation must be carried out with extreme caution and critical diffidence.

In fact starting from data deduced from past experience can give rise to new statical problems, for example of instability, which are possible causes of collapse and disaster not considered in the light of traditional analytical procedures.

Finally analytical methods are particularly inadequate in the case of geo-mechanical problems that are concerned with the stability of complex and randomly formed rock masses.

It is useful to realise that in this case the analyses must be carried out mainly in the visco-plastic range, an area in which analytical procedures are as yet not adequately developed.

In models the structural scheme, within the bounds set by the needs of mathe-matical correlation, is more easily reproduced and gives a greater wealth of detailed information.

In particular models are more suitable for demonstrating the deformation processes up to failure, bringing into play in correct proportions and in their entirety the whole system of load carrying components of the structure.

It is meanwhile curious to note how even today one is led to comment at length on the particular use of elastic models as tools for correlating and checking analytical results, where the model is frequently thought of as a means of reproducing the analytical results rather than the behaviour of the real struc-ture; as if frightened that the model may produce results dissimilar from the analytical calculations rather than actively seeking to obtain a richer and more reliable wealth of information.

Such a conception degrades the model and with it the associated laboratory techniques and reduces the role of the researcher to that of a simple technician.

It is likewise strange to note how the designer too frequently limits his inquiry solely to phenomena that can be studied by analytical procedures.

This attitude is justified for the pure scientist for whom a fact or a pheno-menon can only be considered known and established when verified in exact terms.

The attitude of the structural engineer must of necessity be substantially different for he cannot ignore a phenomenon solely because there is not available an exact and codified method of analysis, nevertheless the less well a phenomenon is understood then the more must considerations of prudence and responsibility lead to a thorough critical examination of the problem.

Faced with such a situation he must start, if necessary, from traditional design methods and make use of all methods of investigation that can provide elements of information, sometimes even simply in qualitative and comparative terms.

In a situation of this type the technique of modelling leaps into the vanguard as a right and proper research tool suitable for reducing the cloud of ignorance surrounding the mechanics of the phenomenon under investigation.

Looked at from another point of view the value of data cannot be measured in terms of its precision but rather in ratio to the help it can provide and the value it acquires in resolving the given problem.

The experimentally deduced statical behaviour of a simply supported beam certainly gives a much more precise picture than results deduced for example from a geomechanical model, but whereas the knowledge gained from the former adds little or nothing to an analytical investigation the latter, despite being coarse and approximate, very frequently represents the only means of investigation for a rock mass.

The choice of an adequate factor of safety must fairly take into account the degree of approximation.

In such a situation it is possible to assign more general duties and objectives to experiments on models than can be obtained from a simple casual investigation. They can become a right and proper research tool capable of providing ideas and consolidating observations required for the realisation of bolder and more advanced construction methods.

Finally it is useful to observe that experimentation on models provides young architects and engineers with a magnificent mental discipline capable of assuring a rich wealth of precious experience, resulting from actual observations, capable of refining the appreciation of static behaviour of the future designer.

Nevertheless the fine thoughts developed here are certainly not original if we give credit to the great Michelangelo who, according to Vasari, said "Models are the most blessed wealth that can be bestowed on one who builds".

Bergamo, Italia, March 1973
Viale Giulio Cesare, 14   **E. Fumagalli**

Contents

1. Principles of Similitude

1.1 Quantities, Magnitudes, Dimensions, Dimensional Systems

According to the principles of mechanics, having acquired the concept of "quantity", if we wish to define the sum of two homogeneous quantities it is necessary to introduce the concept of "magnitude" as the ratio between the quantity under consideration and another one, homogeneous to it, chosen as unity. We define a "dimensional system" as the whole set of derived quantities whose units can be obtained from certain predetermined "fundamental" units. Among the various dimensional systems we mention those of:

geometry, where all the quantities can be derived from one fundamental quantity represented by the length L;

kinematics, where all the quantities can be derived from those of length L and time T;

statics, where all the quantities can be derived from those of length L and force P, or (see 1.4) length L and specific force;

mechanics in general, in which all the quantities can be derived from the above three fundamental ones.

We will limit our consideration to mechanical systems, without looking into other branches of physics.

It is well known that the choice of the three fundamental quantities of mechanics is entirely arbitrary, with the sole condition that they are independent of each other. In fact in mechanics there are at least three separate systems of units in universal use: C.G.S. system, the M.K.S. or Giorgi system and the technical or practical system. These systems are said to be "coherent", inasmuch as they reduce to the minimum the required "parasitic coefficients", in contrast with the situation when the units are chosen arbitrarily without reference amongst themselves (as exemplified by the traditional English system).

It is important to note that the relation by means of which the value y of a certain "derived" quantity is deduced from the values $\bar{x}_1 \dots \bar{x}_q$ of the q fundamental quantities on which it depends, must necessarily be a monomial relation, that is of form:

$$y = K \, \bar{x}_1^{\alpha_1} \cdot \bar{x}_2^{\alpha_2} \dots \bar{x}_q^{\alpha_q} \tag{1}$$

where K is a numerical coefficient, that must be a constant or a function of the ratio between homogeneous physical quantities[1], if the relation between two homogenous quantities of the type y is to remain constant with respect to variations of the units chosen for the values $\bar{x}_1 \ldots \bar{x}_q$.

The notation $[y] = [L^\alpha T^\beta M^\gamma]$ (using the C.G.S. system) was introduced by Maxwell to show the dimensional dependence of the derived quantity on the fundamental ones, and indicates that the value of y is obtained by multiplying together α values of length, β values of times, γ values of mass.

If all the exponents are zero, we have a non-dimensional quantity and its value is defined by a pure number.

In Table 1 are listed some of the more important quantities for our problems, together with their dimensions, in a mechanical system (C.G.S.) and a statical system (L, P).

Table 1

Linear dimensions	L	L
Force or concentrated load P	MLT^{-2}	P
Unit pressure	$ML^{-1} T^{-2}$	PL^{-2}
Specific weight	$ML^{-2} T^{-2}$	PL^{-3}
Modulus of elasticity E	$ML^{-1} T^{-2}$	PL^{-2}
Yield point (or limit of plasticity)	$ML^{-1} T^{-2}$	PL^{-2}

In addition to the above quantities there are some other non-dimensional ones, the values of which are given by pure numbers, for example:

Poisson's ratio v,

the angle of internal friction φ of a material.

1.2 Principle of Homogeneity and Buckingham's Pi Theorem

Let us consider a certain physical phenomenon, involving n quantities $x_1 \ldots x_n$: each of which can be a variable, or a constant, with dimensions non zero, or zero.

The physical law of the phenomenon will be expressed (in implicit form) by an equation:

$$f(x_1 \ldots x_n) = 0 \qquad (2)$$

The "principle of homogeneity" says that, given that f is composed of many terms, these must all have the same physical dimensions: it is clear in fact, that

---

1 If a circle of radius r is drawn on the surface of a sphere of radius R then the area of the surface of the sphere included within the circle is given by:

$$A = 2\pi \left(1 - \cos\frac{r}{R}\right) R^2 = \pi \left(\sin\frac{r}{2R} \middle/ \frac{r}{2R}\right)^2 r^2$$

where the coefficient $K = \pi \left(\sin\frac{r}{2R} \middle/ \frac{r}{2R}\right)^2$ is a function of $r/2R$.

if this is not so, then changing the units of measurement will result in different terms being multiplied by different factors, and equation (2) will not be uniquely satisfied. Further, given that some of the terms of (2) may contain non-monomial expressions (exponentials, logarithms, trigonometric functions, etc.) then these must necessarily be non-dimensional.

Equation (2) can always be put in the form:

$$f\ (\pi_1\ ...\ \pi_n)=0 \tag{2'}$$

where the π_n are the non-dimensional relations corresponding to the x_n (for this purpose it is sufficient to divide by any one of the terms of equation (2)).

It may be that not all the π_n are themselves independent, that is some of them may not say anything new that has not already been said by others; in such a case the number of significant relations is reduced to m where $m<n$.

Buckingham's theorem (the proof of which will be omitted) says that if there are q fundamental quantities then the number m of the independent relations π is given by $m=n-q$, and therefore instead of equations (2') we now have:

$$F\ (\pi_1\ ...\ \pi_m)=0 \tag{3}$$

in which only the significant non-dimensional relations are included. Sometimes we consider separately the case in which some of the quantities $x_1\ ...\ x_n$ are of the same type (for example two or more lengths, two or more forces, yield points and axial failure loads), although it is included in the general theorem. In this latter case we can treat separately the non-dimensional relations, that we will call say χ_i, of all the quantities of this type. Such "form factors" must be contained in F of equation (3) if they have a real influence on the phenomena. We will therefore re-write equation (3) in the following form:

$$F\ (\pi_1\ ...\ \pi_m:\ \chi_2,\chi_2\ ...)=0\ . \tag{3'}$$

This represents, in a more general form, Buckingham's π theorem and this will form the basis of what follows.

In particular the explicit form of equation (3') with respect to π_1 is:

$$\pi_1=\phi\ (\pi_2\ ...\ \pi_m;\ \chi_1,\chi_2\ ...). \tag{3''}$$

For some examples of the application of dimensional analysis see the paper by Pistolesi cited in the bibliography [23].

1.3   General Concepts of Similitude

In equation (3') which, as we have already said, expresses the law of the physical phenomenon under study, some quantities will be "data" of the problem and others will be unknown "functions".

If there is more than one function we will naturally need the same number of independent equations of the type (3') as there are unknowns. Whatever our interest, we can consider the case of a single unknown function, let it be x_1, which from now on will be denoted by y for clarity.

As indicated π_1 has the form (see equation (1)):

$$\pi_1 = \frac{y}{\bar{x}_1^{\alpha_1} \ \dots \ \bar{x}_q^{\alpha_q}} \tag{4}$$

where the $\bar{x}$ are the fundamental quantities.

From (3'') and (4) we have:

$$y = \bar{x}_1^{\alpha_1} \ \dots \ \bar{x}^{\alpha_q} \phi \left(\pi_2, \pi_3 \ \dots \ \pi_m ; \chi_1, \chi_2 \ \dots \right). \tag{5}$$

This is the usual form of the physical law in dimensional analysis. Two systems are said to be physically similar when we can determine a unique correspondence between all points of the two systems and when the physical quantities have a constant relationship at corresponding points.

In problems that involve the shape of the body, the previous definition necessarily requires geometric similarity.

The idea of dimensional similarity shows that when passing from one mechanical system to a "similar" one the constant ϕ of equation (5) (since its mathematical form cannot be changed) takes on the same value, and therefore the $(m-1)$ functions π assume the same values as also do the coefficients of the type χ.

We can now write equation (5) relative either to the prototype or to the supposed model indicating model quantities by a superscript dash. We can then divide the prototype quantities by the model quantities and if similitude is respected the resulting ϕ are equal and will cancel out.

It then follows that:

$$y = y' \left(\frac{\bar{x}_1}{\bar{x}_1'} \right)^{\alpha_1} \ \dots \ \left(\frac{\bar{x}_q}{\bar{x}_q'} \right)^{\alpha_q} = \Omega \, y' \tag{6}$$

that is from the value y' derived from the model we can pass to that of y of the prototype by means of a simple scale factor Ω, the value of this latter being determined from the scale relations $\bar{x}/\bar{x}'$ relative to the fundamental quantities; this relation obviously being fixed by similitude when we design the model.

Similarity can also be guaranteed even though we do not explicitly know the form of ϕ in equation (5) provided that we do not forget in equation (2) any significant quantities of the phenomenon under study and that we can guarantee the preservation of all the non-dimensional relations of equation (5) relative to the derived quantities and of all the form factors.

1.4 Similarity in Problems of Statics

It is well known that in problems of statics we require two independent quantities which can be freely chosen and to which all other derived quantities must be referred.

For convenience we choose length as our first fundamental quantity, fixing the relative scale relation between prototype and model as $\lambda = L/L'$, and as second fundamental quantity we choose that of specific force fixing the relative scale as $\zeta = \sigma/\sigma'$.

In particular we note that the scale of the length λ must also be valid for the displacements, be they absolute or relative and that the scale ζ must be valid for all quantities having the dimensions of specific force, for example:

$$\zeta = \frac{\sigma}{\sigma'} = \frac{E}{E'} = \frac{\sigma_{uc}}{\sigma'_{uc}} = \frac{\sigma_{ut}}{\sigma'_{ut}} \dots \text{etc.} \tag{7}$$

Where E represents the modulus of elasticity or Youngs modulus, σ_{uc} the ultimate compressive strength, σ_{ut} the ultimate tensile strength etc.

A material that faithfully reproduces to this scale the mechanical properties of a prototype material is defined as a material that has a "degree of efficiency ζ" with respect to the prototype material [21].

Let us consider now some derived quantities, for example the specific weight γ of a solid or of a liquid. In this case we observe that the relative dimensions expressed in the fundamental quantities are:

$$\gamma = \sigma \cdot L^{-1},$$

from which we find that the relative scale of reproduction is equal to:

$$\rho = \frac{\gamma}{\gamma'} = \left(\frac{\sigma \cdot L^{-1}}{\sigma' \cdot L'^{-1}} \right) = \zeta \cdot \lambda^{-1}. \tag{8}$$

For a concentrated force on the other hand:

$$P = \sigma \cdot L^2$$

and the scale relation becomes:

$$\psi = \frac{P}{P'} = \frac{\sigma \cdot L^2}{\sigma' \cdot L'^2} = \zeta \cdot \lambda^2. \tag{9}$$

From equations (8) and (9) are derived the two fundamental relations between the scales of the quantities that normally occur in problems of statics:

$$\zeta = \rho \cdot \lambda, \tag{10}$$

$$\psi = \rho \cdot \lambda^3. \tag{11}$$

It is also useful to consider the strains:

$$\varepsilon = \lim_{L \to 0} \frac{\Delta L}{L},$$

from which non-dimensional relation we have that:

$$\varepsilon = \varepsilon'. \tag{12}$$

It is significant that the relative strains at geometrically similar points remain unchanged when passing from the prototype to the model. The same applies to Poisson's ratio (a non-dimensional relation between the induced normal strain and the initial direct strain) as also for the angles of friction, be they of internal friction of a material or relative to potential sliding surfaces.

In practice, referring to equation (10), modelling in the field of statics can be sub-divided into two fundamental categories:

1. Models in which only the scale of the length $\lambda > 1$, while that of the stress $\zeta = 1$. In this case we can use the material of the prototype for the model or at least a model material with properties equal to those of the prototype.

Such models can also be used for tests carried to failure beyond the elastic region.

The limits of scale reduction for this category of models are frequently fixed by the practical difficulty of complying with the scale relations for body forces according to equation (10).

2. Models in which both the scale of the length λ, and the scale of the specific forces ζ are > 1. We can further sub-divide this category into two parts dealing with substantially different areas of application:

a) models for use only in the linear elastic region;
b) models for use beyond the elastic region.

For the category 1. it follows from equation (10) that:

$$\lambda = \frac{1}{\rho},$$

that is to say:

$$\gamma' = \gamma \cdot \lambda.$$

Practical models of this type are limited in general to those requiring relatively small scale reduction factors λ, such as structural elements in reinforced concrete (beams, portals, floor slabs etc.).

It is of interest, however, that recently there have been remarkable developments in the application of models to nuclear reactor pressure vessels in prestressed concrete.

Models of the category 2. a) are used only in the elastic region and are not necessarily required to satisfy equation (12); thus one can vary at will the intensity of the loads and, therefore, the deformations proportional to these loads, provided that one remains within the linear elastic range of the material.

In such cases it is necessary to determine the load and deformation relationships between prototype and model. The experimental model now acts like a theoretical method of calculation that takes account of all the complex conditions of bond and structural rigidity in the region of linear elastic behaviour.

Thus for example for an investigation of a large reinforced concrete structure the elastic model furnishes the resulting forces and displacements for the individual structural elements, on the hypothesis of sections acting entirely elastically (not plastically), exactly according to the assumptions of an overall theoretical analysis always providing that we remain within the linear elastic region.

The only material properties we require to know or determine in this case are the elastic moduli of the materials of the model and prototype, thus fixing the scale $\zeta = \frac{E}{E'}$.

In order to facilitate accurate measurement of the deformations, it is often convenient to increase the intensity of the applied loads (beyond that required for scale deformations). This is only permissible for deformations within the elastic range, when we use equations (10) and (11) to interpret the results.

The materials used for the models must have linear load deformation characteristics over a very wide range.

Models of category 2. b) are usable up to rupture [22]. This type of modelling represents in practice the more suitable method for investigations beyond the normal limits of analytical methods. It also provides information on the internal behaviour of the structure beyond the elastic limit, in the elasto-plasto-viscous region and ultimately allows us to determine, through tests taken to failure, the overall factor of safety.

Body forces can be accomodated, according to equation (10), by using materials with a sufficiently large degree of efficiency ζ.

Choosing in particular $\zeta = \lambda$, we can reproduce interesting rock systems with frequent discontinuities (geomechanical models). Artificial methods of reproducing self weight by means of internally anchored loading bars (see Chapter 3) cannot be used in this type of model because the loading bars tend to restrain sliding on potential slip plains, especially in regions of large deformation.

Taking a more general look at the required characteristics of a material to be used to rupture, we note that the Mohr envelopes of the model and prototype materials (Fig. 1a) [11] must be geometrically similar with respect to the origin of the axes according to the relations:

$$\sigma = \zeta\, \sigma' \qquad \tau = \zeta\, \tau'.$$

Analogously the stress strain curves of the prototype and model materials must correspond according to the two relations:

$$\sigma = \zeta\, \sigma' \qquad \varepsilon = \varepsilon'.$$

The two curves are linked by an orthogonal relation (Fig. 1b). This latter diagram should logically take into account the elastic and anelastic deformation phenomena including those that depend on the variable time: flow, viscosity, hysteresis, relaxation etc. If in particular the deformation processes depend on the variable time, the system becomes a dynamic one and passes from the field of statics to that of mechanics (1.1). In this case there will be three independent quantities.

However we find that as long as we operate on an equipotential surface, of acceleration g, the scale of the new independent quantity must in every case satisfy the following relation:

$$G = \frac{g}{g'} = \frac{L\ T^{-2}}{L'\ T'^{-2}} = \lambda\, t^{-2} = 1,$$

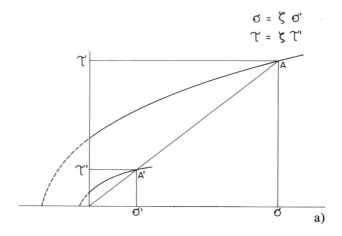

$$\sigma = \zeta \, \sigma'$$
$$\tau = \zeta \, \tau'$$

a)

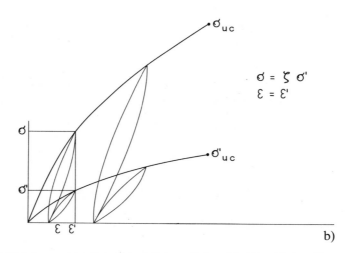

$$\sigma = \zeta \, \sigma'$$
$$\varepsilon = \varepsilon'$$

b)

Fig. 1. a) Mohr's representation of strength curves of model and prototype, b) stress-strain relation of prototype and model

from which we find that the time scale between model and prototype is equal to:

$$t = \lambda^{\frac{1}{2}} \, . \tag{13}$$

In particular for experiments on geomechanical models we should take into consideration the deformation process of "creep".

At the present time sufficient experimental evidence does not exist to enable us to put forward a reliable hypothesis on the laws that govern creep. There does not even exist a reliable and rigorous method of experimentation for the prototype material.

Nevertheless let us examine the behaviour of a cylindrical sample of a hypothetical rock, the matrix of which can be considered isotropic and homogeneous, loading it in a triaxial cell with a sufficiently high isotropic pressure such that its cohesive strength is negligible by comparison. For sufficiently compact rock the discontinuities and the planes of preferred rupture do not influence the deformability and rupture of the material.

We now apply a deviatoric component so as to induce appreciable amounts of elastic and creep deformation, without reaching rupture.

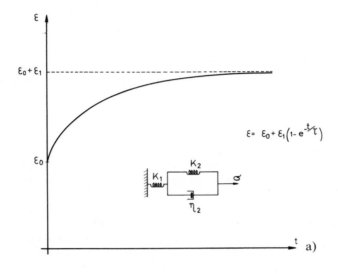

$$\varepsilon = \varepsilon_0 + \varepsilon_1 \left(1 - e^{-t/\tau} \right)$$

a)

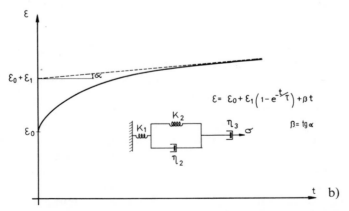

$$\varepsilon = \varepsilon_0 + \varepsilon_1 \left(1 - e^{-t/\tau} \right) + \beta t$$

$$\beta = \text{tg}\,\alpha$$

b)

Fig. 2. a) Time dependent deformation process according to the Kelvin rheological model, b) time dependent deformation process according to the Burgers rheological model

As long as the deforming material attains a new condition of equilibrium without fracturing, the generalised Kelvin solid gives a reasonable representation of the phenomenon.

The rheological model is shown in Fig. 2a, where:

K_1 is the constant of the series spring,

K_2 is the elastic constant of the parallel spring,

η_2 is the coefficient of the viscosity of the dashpot.

This representation produces a deformation behaviour with time, given by the relation [14]:

$$\varepsilon = \varepsilon_0 + \varepsilon_1 \left(1 - e^{-t/\tau}\right),$$

in which:

ε_0 is the elastic deformation,

$\varepsilon_1 = \sigma/K_2$ represents the creep deformation due to a deviator stress σ for t ∞,

$\tau = K_2/\eta_2$ is a constant of time,

$\varepsilon = \varepsilon_0 + \varepsilon_1$ is the deformation asymptotic to a new equilibrium attained at time t ∞.

We now imagine that the deviator load is increased sufficiently to produce a slow rupture of the sample. Since in the sample all the planes offer equal resistance to failure, failure (according to Mohr's representation) will occur as a large number of sliding surfaces defined by a two fold system of opposed cones with their vertices along the axes of the sample and with internal angles equal to $\pm(45° + \varphi/2)$, where φ is the angle of internal friction of the material.

In practice, due to the effect of tangential constraint between the metal loading platens and the ends of the sample, the sample deformations acquire the classical barrel form which governs the local increase of the applied isotropic stress tensor.

This deformation process is more closely reproduced by the Burgers solid (Fig. 2b) which, compared with the previous rheological model, contains an additional series dashpot of constant η_3.

The deformation behaviour with time of this model is given by:

$$\varepsilon = \varepsilon_0 + \varepsilon_1 \left(1 - e^{-t/\tau}\right) + \beta\, t,$$

where β represents the deformation velocity for $t\infty$.

The inclined asymptote provides one with a typically viscous region of deformation along which the material moves to rupture. It is valid naturally in the case of small deformations but for large deformations variations in the loaded sections occur that substantially modify the stress regime, so that for very high values of the deviator stress the material moves to failure without permitting the inclined asymptote to be followed, even in a transitory manner. It may be noted that through similar tests it is possible to include the behaviour of creep in easily interpreted elementary rheological schemes. From these one can then deduce the values of the coefficients of the interpolation curves of the experimental data, thus providing intrinsic parameters of the material that are of fundamental importance in trying to establish possible mathematical or numerical analyses (see for example the method of finite elements), besides also being necessary for defining the parameters for similitude required in designing models.

Tests of this type are carried out at Ismes only on model materials that have high ratios of efficiency ζ (materials for geomechanical models), with highly encouraging results. When we wish to carry out the above tests on real rocks the situation is evidently (for practical reasons only) more complex and difficult and one is obliged to work at very high confining pressures.

Previously, in the years between 1920—30, Ros explored the processes of deformation and rupture in similar conditions, albeit for less ambitious reasons.

Today, as we are doing, it is worth re-examining the problem in a more detailed way to determine the parameters of interest that control the deformation processes. Nevertheless, while mathematical analyses based on elementary rheological models allow ample and impressive analytical treatments, there does not yet exist a sufficiently large amount of experimental information to include the behaviour of a particular rock in a particular type of rheological scheme. For the present it appears logical to abandon a too rigid attitude and to re-examine the problem within less ambitious bounds.

We limit ourselves therefore in the model to reproducing the values of elastic and anelastic deformations for total completion of each deformation process. By this device the problem is reduced solely to that of reproducing an equivalent static system.

To put order into the ideas that interest us and to obtain a possible idealisation necessary for reproducing the deformation processes in the model, we consider a compression test on a sample of rock sufficiently large as to be reasonably representative of the material.

We imagine that we carry out the test in such a way that for each load there is a point on the stress deformation diagram that represents complete stabilisation of the material, that is to say for total completion of all deformations for that state of stress. This point will be determined after the execution of a sufficient number of cycles of adjustment and application of the stress over a sufficiently long time.

If we determine a sufficient number of values of total deformation for gradually increasing stresses σ_A, σ_B, σ_C (Fig. 3) it is possible to construct a stress deformation curve relative to a state of complete stabilisation of the material.

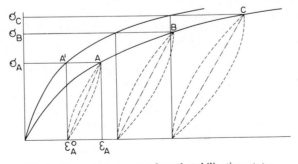

Fig. 3. Stress-strain curve of total stabilization states

Suppose we attain but do not exceed the specific stress σ_A and reach stabilisation of the material for this stress. If we then execute compressive cycles between σ_A and σ, we obtain cycles that trace out a hysteresis loop which, although they are irreversible, are nevertheless repeatable; provided that we agree on a suitable method of test.

We can then define:

the modulus of deformability for repeated cycles in relatively rapid time as:

$$E_r = \frac{\sigma_A}{\varepsilon_A - \varepsilon_A^0}, \tag{14}$$

the secant modules of total deformability to completion of the deformation as:

$$E_s = \frac{\sigma_A}{\varepsilon_A} \tag{15}$$

the value ε_A° of anelastic deformation which is evidently composed of:

i) creep of the rock matrix;

ii) visco-plastic adjustment connected with the closure or sliding of the internal discontinuities of the rock. If the contacting surfaces are in sound rock then the adjustments are of a mainly plastic nature; whilst in the presence of infillings of peat or moist clay the adjustments are of a mainly viscous nature.

References

[1] Arountiounian, N.: Applications de la théorie du fluage. Paris: Eyrolles. 1957.

[2] Beaujoint, N.: Similitude et théorie des modèles. Rilem, Bulletin Nr. 7, 14—39, Paris (June 1960).

[3] Bertrand, J.: Note sur la similitude en mécanique. Journal de l'Ecole Polytechnique, t. XIX, Cahier XXXII, 1848.

[4] Bonvalet, M.: La similitude des maquettes en élasticimétrie. Exemples d'application à deux ouvrages d'art. Mémoire du G.A.M.A.C. (February 1954).

[5] Borges, J. F.: Statistical Theories of Structural Similitude. Rilem, Bulletin Nr. 7, 59—67, Paris (June 1960).

[6] Borges, J. F., Arga, L. J.: Crack and Deformation Similitude in Reinforced Concrete. Rilem Bulletin Nr. 7, 79—90, Paris (June 1960).

[7] Bridgman, P. W.: Dimensional Analysis. Yale University Press. 1952.

[8] Duncan, W. J.: Physical Similarity and Dimensional Analysis. London: Arnold 1953.

[9] Esnault Pelterie, R.: L'analyse dimensionnelle. Paris: Gauthier-Villars; Lausanne: Ed. Rouge. 1946.

[10] Fumagalli, E.: Model Simulation of Rock Mechanics Problems. Rock Mechanics in Engineering Practice (Stagg-Zienkiewicz, eds.). London: J. Wiley. 1968.

[11] Fumagalli, E.: Caratteristiche di resistenza dei conglomerati cementizi per stati di compressione pluriassiale. Ismes, Bulletin Nr. 30 (October 1965).

[12] Gabrielli, G.: Le esperienze su modelli e le leggi di similitudine. Ingegneria Meccanica IV, 1955, 2.

[13] Ghaswala, S. K.: Model and Analogies in Structural Engineering. Civil Engineering and Public Works Review (January—June, 1952).

[14] Jaeger, J. C., Cook, N. G. W.: Fundamentals of Rock Mechanics. London: Methuen. 1969.

[15] Jasiewicz, J.: Applications of Dimensional Analysis Methods to Civil Engineering Problems. Civil Engineering and Public Works Review, London, Vol. 58, Nr. 686, 1125—1128 (September 1953); Nr. 687, 1270—1280 (October 1963); Nr. 688, 1409—1410 (November 1963); Nr. 689, 1547—1549 (December 1963).

[16] Mattock, A. H.: Structural Model Testing — Theory and Applications. Journal, PCA Research and Development Laboratories, Vol. 4, Nr. 3, 12—13 (September 1962).

[17] McHenry, D., Karni, J.: Strength of Concrete under Combined Tensile and Compressive Stress. ACI Journal, Proceedings, Vol. 54, Nr. 10, 829—840 (April 1958).

[18] Murphy, G.: Similitude in Engineering. New York: Ronald Press. 1950.

[19] Nowacki, W.: Théorie du fluage. Paris: Eyrolles. 1965.

[20] Oberti, G.: Sulla valutazione del coefficiente globale di sicurezza di una struttura mediante esperienze su modelli. Ismes, Bulletin Nr. 2 (June 1954).

[21] Oberti, G.: Large Scale Model Testing of Structures outside the Elastic Limit. Ismes, Bulletin Nr. 12 (April 1959).

[22] Oberti, G., Lauletta, E.: Evaluation Criteria for Factors of Safety. Model Test Results. Ismes, Bulletin Nr. 25 (October 1964).

[23] Pistolesi, E.: Omogeneità, similitudine, modelli. Questioni di matematica applicata, Collana CNR. Bologna: Zanichelli. 1940.

[24] Pletta, D. H., Frederick, D.: Experimental Analysis. Proceedings ASCE, Vol. 79 (July 1953).

[25] Somerville, G.: The Behaviour and Simulation of Concrete. Presented to the Joint British Committee for Stress Analysis at a Meeting entitled: The Behaviour and Simulation of Engineering Materials. University College, London (January 2, 1967).

[26] Templin, R. L.: Tests of Engineering Structures and their Models. Transactions, ASCE, Vol. 102, 1211—1225 (1937).

[27] Wilbur, J. B., Norris, C. H.: Structural Model Analysis. Handbook of Experimental Stress Analysis. New York: J. Wiley. 1950.

[28] Wright, J.: Use of Models in Structural Analysis. Engineering (London), Vol. 175, Nr. 4558, 110—712 (June 5, 1953); Nr. 4559, 741—743 (June 12, 1953).

[29] Zia, P., Van Horn, D. A., White, R. N.: Principles of Model Analysis. Models of Concrete Structures, Special Publication, Nr. 24. Detroit: American Concrete Institute. 1970.

2. Physico-Mechanical Properties to Be Reproduced in the Model Materials

2.1 Introduction

In the preceding chapter we discussed the parameters of similitude that must be respected by the model materials. Naturally it is worth limiting these parameters to those that have a direct influence on the phenomenon to be studied. For example, for a static investigation limited to the elastic region we have seen that it is sufficient not to produce deformations beyond the linearly elastic range in the course of the tests. In this case the only value that needs to be considered is the ratio ζ between the Young's moduli, that represents the "ratio of efficiency" of the model material.

Again for experiments up to failure on statistically determinate structures it is only necessary to apply the ratio ζ to the Young's moduli and ultimate strenghts (tension, compression), regardless of the complex visco-plastic phenomenon that may intervene between yield and failure. We must bear in mind, however, that the problems of greatest interest in experiments to failure are those concerning hyperstatic structures, especially when supported on foundations of varying flexibility.

In such cases it is obviously necessary to take into consideration the entire range of the physico-mechanical properties of the materials, including therein, if necessary, the processes of creep as considered schematically in chapter 1.4.

On this point, it is also a good rule that in choosing materials we should not limit ourselves only to consider the reproduction of the theoretical physico-mechanical properties, as outlined in the preceding chapter, but that we should take into account when possible the physical structure, seeking to reproduce in this way the complex nature of the material. For example the type of material most suitable for reproducing a cemented conglomerate is itself normally a conglomerate.

We now proceed to a detailed examination of materials, dividing them into two fundamental categories.

2.2   Materials for Experiments in the Elastic Range

A variety of materials are known that are suitable for models limited to elastic deformations.

Among these we can list: celluloid, rubber and, in general, a wide range of thermo-plastic and thermo-setting resins that are now available [5].

The greatest limitation of all of these materials is that they have Poisson's ratios varying between 0.35 and 0.48, these being substantially higher than that of concrete ($v = 0.2$). This has a significant effect on the results for massive three-dimensional models and for models highly constrained hyperstatically internally or externally.

In some cases the materials can be obtained ready for use in sheets of different thicknesses, which can be glued with suitable adhesives; in this category we have celluloid and acrylic, polystyrene and phenolic resins.

In other cases the model is constructed by casting the material in suitable moulds where hardening occurs by polymerisation. Materials of this type include polyester and silicone resins and in particular a wide range of epoxy resins. The use of these latter covers the majority of applications for experimental models. In fact, they offer important advantages such as:

i) limited development of heat of polymerisation that assures a more homogeneous hardening process. Only when we have a homogeneous maturation in the cast can we be sure of obtaining a constant elastic modulus throughout the mass. This is particularly important in models of varying thickness.

ii) reduced shrinkage, which results in a significant reduction in internal stress. This point is of fundamental importance in photoelastic models, and also when moulding models of large thickness in which case the internal stresses can cause fissuring and fracture in the model.

iii) the possibility of grading the deformability properties by varying at will the quantity of hardener, or more generally by using the many types of these resins produced by industry.

In Table 2 there are shown the particular properties of some of the more useful resins. This information is taken from a publication by Roll [21].

It is obviously not possible in the present work to examine in detail the properties of all the thermo-setting resins available for casting models. There is widespread industrial interest in these resins, which are finding there an advanced field of application.

Cast models should, however, only be used in well organised laboratories where there is available a high degree of technical skill and are better avoided under less favourable circumstances in which it might be necessary to improvise. Otherwise it is perhaps preferable to construct models from glued sheets, especially for plane models.

In the latter case, celluloid is a particularly suitable material since joints in the sheets can be made by self adhesive methods. In fact, by applying acetone

Table 2. *Properties of Some Plastics Suitable for Structural Models*

Plastic	Thermal Characteristic	Available Shapes	Modulus of Elasticity in kg/cm^{-2}	Poisson's Ratio	Tensile Strength in kg/cm^2	Specific Gravity	Softening Temperature, in degrees Farenheit	Coefficient of Expansion in cm per cm per degree Farenheit	Jointing Characteristics
Cellulose Nitrate (Celluloid)	thermo-plastic	sheets, rods, and tubes	$20-25 \cdot 10^3$	0.40—0.42	20— 40	1.35—1.70	212	$7- 9 \cdot 10^{-5}$	can be cemented with solvent cements and ethyl acetate (acetone)
Cellulose Acetates (Plasticele)	thermo-plastic	sheets, rods, and tubes	$18-20 \cdot 10^3$	0.40	300—350	1.32—3.33	500	$8-10 \cdot 10^{-5}$	can be cemented with solvent cements and ethyl acetate (acetone)
Methyl Methacrylates (Plexiglas, Leucite, Perspex)	thermo-plastic	sheets, rods, and tubes	$30-35 \cdot 10^3$	0.35—0.38	500—750	1.18—1.20	250—320	$4- 5 \cdot 10^{-5}$	can be cemented with commercial adhesives or a solution of the plastic and chloroform
Polyvinyl Chlorides (Boltaron)	thermo-plastic	sheets, rods, and tubes	$30 \cdot 10^3$	0.38	700 at yield	1.43	220	$3.7 \cdot 10^{-5}$	can be welded with PVC rods or cemented with epoxy cements
Polyethylenes (Alkathene)	thermo-plastic moulding powders	sheets, rods, tubes,	$2- 3 \cdot 10^3$	0.45—0.50	75—100	0.92—0.94	212	$9 \cdot 10^{-5}$	can be welded but difficult to cement

Polyester Resins (Marco, Palatal)	thermo-setting	casting resins	$20-30 \cdot 10^3$	0.33—0.35	350—400	1.20—1.30	180	$3-6 \cdot 10^{-5}$ (solvents are acetone, and cellosolve)
Epoxy Resins (Epon, Araldite)	thermo-setting	casting resins	$30-50 \cdot 10^3$	0.33—0.35	500—700	1.20	—	$3 \cdot 10^{-5}$ can be cemented with epoxy cements

Table 3. *Composition and Properties of some Epoxy Resin Mixes with Added Fillers*

Resin base % by weight		Filler materials % by weight				Physico-mechanical properties	
Epoxy resin (Araldite M)	Hardener	Cork dust	Aluminium podwer	Polystyrene granules	Silica sand 0.5—1 mm	Elastic Modulus $E = \mathrm{kg/cm}^{-2}$	Poisson's Ratio
83.3	16.7	—	—	—	—	26,000— 32,000	0.38
71.5	14.3	—	14.2	—	—	25,000— 30,000	0.28
38.5	7.7	38.4	15.4	—	—	5,000— 7,000	0.27
23.7	4.8	—	—	71.5	—	18,000— 25,000	0.34
15.4	3.0	—	—	4.6	77.0	80,000— 90,000	0.27
9.8	2.0	—	—	—	88.2	140,000—160,000	0.26

(solvent) to the surfaces to be joined self produced welding and strength is obtained by simply pressing the surfaces together.

Perspex (plexiglas) sheets can also be easily and securely glued with appropriate adhesives and have the advantage over celluloid of being less inflammable and more stable with respect to variations of ambient temperature and humidity.

However, apart from the technological difficulties of preparation, casting resins afford a further wide field of application: they can be filled with inert materials dispersed homogeneously through the mass. As inert fillers we can use: silica sand, powdered magnetite, iron, aluminium, cork, litharge, lead shot, polystyrene granules, etc. The presence of fillers reduces the temperature rise due to the generation of heat of polymerisation. It also reduces shrinkage and the development of internal stress; it can also be used to increase the density of the material for self weight tests and to modify, within fairly wide limits, its modulus of elasticity.

The addition of aluminium powder in particular increases the thermal conductivity, facilitating the dispersion of heat produced by electrical resistance strain gauges that usually have to be applied to the surface of these models. The generation of heat by electrical resistance strain gauges applied to the resin surfaces frequently produces difficulties in reading the gauges due to the phenomenon of continous drift consequent on the local temperature rise.

Again, using suitable fillings (powdered cork, sand, polystyrene), we can reduce Poisson's ratio to values closer to those of concrete or steel. Table 3 shows details of the elastic moduli and Poisson's ratios resulting from a range of epoxy resin mixes with varying additions of fillers.

2.3   Materials for Experiments up to Failure

The other category of materials is represented by mortars made with cementaceous binders together with fillers (aggregates) or additives in suitable proportions.

As binders we can use: cement, plaster, lime and synthetic resins and as fillers, either granular or as dust: sand, pumice stone, lead shot, magnetite, barytes, cork, polystyrene granules, litharge (PbO), red lead (Pb_3O_4), etc.

We can also use natural additives such as bentonite and diatomite or industrial additives to improve the plasticity and workability of the mortars.

This category includes all the materials that can be used for experiments taken to failure in accordance with equation (10). They also normally produce a rheological behaviour similar to that of concrete, especially when cement and granular fillers are used.

By a suitable choice of the proportions of the components we can adequately reproduce the basic relations between modulus of elasticity, ultimate strength in tension and compression, as also the stress strain curve up to failure.

For obvious reasons of economy, we will limit ourselves in the present work to an examination of those materials that have been tested and found suitable for the construction of models.

2.3.1 Microconcrete

For models reproduced with a scale factor $\zeta = 1$ the material used should, in theory, be the same as that of the prototype (concrete).

In practice, since the materials are normally used in small thicknesses, it is necessary to make a suitable reduction of the dimensions of the aggregate (microconcrete).

The resulting increase in the specific surface of the cemented particles necessitates an increase in the proportions of cement and water. This is turn results, for similar conditions, in an increased rate of shrinkage of the order of 2 to 3 times compared with that of a normal concrete. This can also lead, in constrained structural elements, to proportionately higher states of internal stress.

With the exception of these observations there is little to add with specific reference to microconcrete that is not already known from the normal techniques of concrete preparation.

We can only try to carry out the curing of the model in suitably conditioned surroundings and, in order to avoid internal stresses, resort to "the technique of humid models" in which the model is coated with an impermeable varnish. This technique will be discussed in more detail in section 2.3.3.

The choice of D_{max}, the maximum dimension of the aggregate, is determined both by the dimensions of the structural sections and the space between the reinforcement, and by the gauge length of the strain gauges that are to be used, noting that this latter should normally be greater than $4\, D_{max}$.

For the best compaction of the casts we advise the use of aggregate gradings of the cubic or else the Bolomey type:

$$p\% = A + (100 - A)\ \sqrt{\dfrac{d}{D_{max}}}$$

where A has a value between 15 and 20 compared with 8 to 12 for a normal concrete. Also the use of vibrators and addition of plasticisers is recommended for increasing the workability and improving compaction.

Obviously the experiments cannot be started for at least 28 days after casting, after which the material will normally have attained 85 to 90% of its final properties. Control tests during the experiments allow compensation to be made for small changes of properties with time.

In all other respects the properties of a normal concrete, with regard to normal variability, are also applicable to microconcrete.

For an accurate reproduction of the ratio E/σ_{uc} it is advisable, in every case, to make a series of trial mixes varying basically the proportions of cement and choosing the most suitable mix.

If we want to increase the weight of the material to reproduce body forces according to equation (10) we can use heavy aggregates such as thick iron stampings (excluding filings and thin sheet) and magnetite sand.

We must bear in mind that the maximum density attainable is 4.5 to 5 ton/m^3 ($\rho_{max} \geq {}^1\!/_2$) and that the use of such aggregates leads, in equivalent situations, to a not inconsiderable increase of mechanical properties.

2*

2.3.2 Mixtures Based on Plaster and Diatomite

Mixtures of plaster and water with varying amounts of diatomite are particularly suitable for a wide range of applications in the elastic field. Compared with resins they have the advantage of having a Poisson's ratio nearer to that of concrete and by varying the proportions one can produce materials with elastic moduli ranging from 5 to 100×10^3 kg/cm^2.

The hardening of plaster occurs through transformation of $CaSO_4 \cdot \frac{1}{2} H_2O$ (hemihydrate) into $CaSO_4 \cdot 2H_2O$ (bihydrate). This is normally brought about by dispersing the plaster in water during the mixing process. The setting time is increased and the final mechanical properties of the plaster greatly reduced by exceeding the required proportions of water. However, when an excess of water is added beyond a certain limit we find the phenomenon of separation among the components occurs.

It is worth while in this case to add diatomite or Kieselguhr to the mix. This is a soft porous deposit of the fossilised skeletons of diatoms having a very high specific surface greater than 1.5×10^7 cm^2/kg. Diatomite is readily found occurring naturally in large sedimentary deposits. Used as an inert filler it absorbs the excess water and reduces the mechanical properties of the material.

Plaster mixes have a very variable rate of hardening, this being mainly governed by the water/plaster ratio (W/P). For mixtures with low W/P ratios it is sometimes useful to use retardants. For a more extensive treatment of this subject the reader is referred to the systematic investigations by Raphael [17].

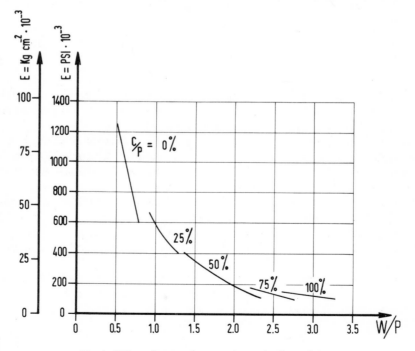

Fig. 4. Effect of water-plaster ratio on elastic modulus

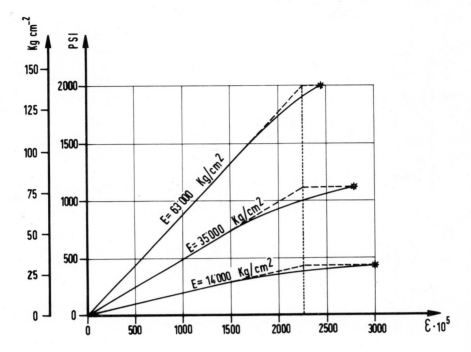

Fig. 5. Ultimate compressive strain

From the graph in Fig. 4 we see that as the W/P ratio of the mixture varies between 0.5 and 0.8 the modulus of elasticity varies approximately between 100 and 50×10^3 kg/cm². With the addition of diatomite in ratios of 25, 50, 75 and 100% by weight with respect to the plaster and varying the W/P ratio between 0.9 and 3.3 we can cover the entire range of moduli of elasticity down to about 8,000 kg/cm².

The stress-strain curves in Fig. 5 show that the plaster based materials have particularly high limit of proportionality and are thus highly suitable materials for elastic models.

From Fig. 5 we further see that, to a first approximation, the limits of elastic deformability are constant irrespective of the elastic modulus. Also the graph of modulus of elasticity/ultimate compressive strength (Fig. 6) is to a good approximation, a straight line given by the relation:

$$\frac{E}{\sigma_{uc}} \triangleq 450$$

However, these materials have an ultimate compressive strength approximately twice that of a normal concrete when compared on the appropriate scale.

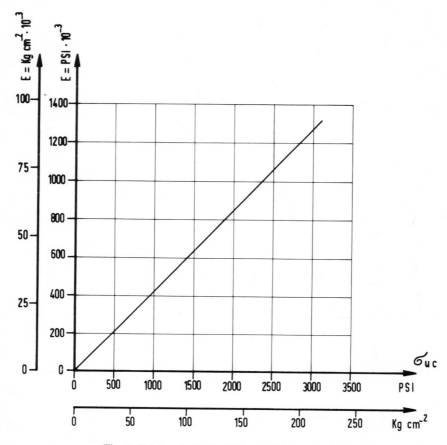

Fig. 6. Relation between elastic modulus and strength

It is nevertheless possible, to a first approximation, to use plaster materials for tests to failure, resorting to a distorted similitude in which the strains in the model and prototype are compared by the relation:

$$\varepsilon = K \cdot \varepsilon'$$

in which $K < 1$ (in the specific case $\simeq 0.6$).

However, such usage should only be considered valid to a first approximation for the following reasons:

i) the behaviour of the material does not adequately reproduce the deformation processes, in the plasto-viscous range, of a normal concrete.

ii) non-uniform drying out of the material induces local variations of properties that can interfere with the processes of fissuration in a manner that is not easily controlled. Only the techniques used at the Laboratorio Nacional de Engenharia Civil in Lisbon (see Chapter 6, Fig. 90) can to a certain extent avoid this problem.

iii) the ratio between ultimate tensile strength and ultimate compressive strength $\sigma_{ut}/\sigma_{uc} = {}^1/_5$th to ${}^1/_6$th is too high (Fig. 7) and is about twice the true value of a normal concrete.

At rupture these materials can reproduce, to a first approximation, concretes of great age and high strength which have very high elastic limits and brittle failure characteristics.

An advantage of these materials is that, provided that they are well dried out, we can use electrical resistance strain gauges on them.

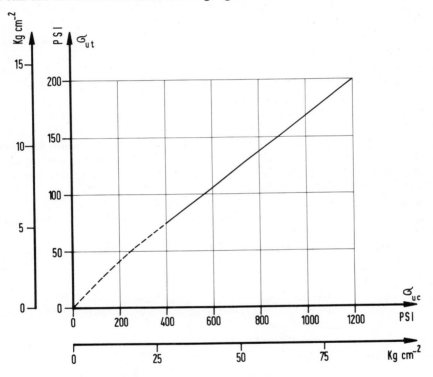

Fig. 7. Relation between tensile and compressive strength

2.3.3 Mortars of Cement and Pumice Stone

Pumice stone and, in general, pozzolanic tuffs have been used as aggregates by Oberti at the Politechnico di Milano since 1936 and are used on a large scale by Ismes at Bergamo [5]. Mortars with pumice should be considered as microconcretes in which conventional stone aggregates are replaced by granulated pumice stone, a porous rock, fundamentally siliceous, of volcanic origin.

In this case the aggregate, relatively weak because of its porosity, appreciably reduces the mechanical properties of the mortars, retaining however the parameters of similitude with concrete so as to produce a ratio of efficiency reduced in the range $3 < \zeta < 15$.

With the pumice-cement mixtures we can in fact cover with good similitude the whole range of elastic moduli within the following limits

$$20 \times 10^3 < E < 100 \times 10^3 \ \text{kg/cm}^2$$

Substituting a fraction of the fine aggregate with conventional sand, in gradually increasing amounts, allows us to extend the range of ζ values that we can cover from 3 to 1.

Because of the high porosity of pumice and the fact that it is hygroscopic, the first stage in the preparation of the mix must be to determine the moisture content of the pumice by heating a sample of it in an oven. The mix is prepared in a vertical mixing machine in small quantities of 30 to 50 litres adding in prescribed order the pumice, the cement, any other additives and then finally dispersing the water during the mixing process to ensure homogoeneous hydration of the cement before the pumice absorbs the excess water.

Casting must be effected slowly but continuously without interruption taking, for example, for the vertical blocks of dam models that can exceed 2 m in height, 10 to 12 hours of working time. Vibration has little effect, the best results for homogeneity and compaction are obtained using a hand pestle with a rubber head, paying attention to ensure complete placing along the surface of the formwork and around the steel reinforcement and loading rods.

Because of the large variety of possible mixtures we will limit ourselves to an examination of some representative materials that have been used for models of dams.

To illustrate materials used to reproduce concrete we have:

Mixture type A

Composition   1,000 litres Lipari pumice (3—7 mm fraction)
              300 kg powdered limestone (fraction between 300 and 1,000 mesh/cm²)
              20 kg diatomite
              3 kg bentonite

respectively in three proportions:

		A_1	A_2	A_3
Portland cement	kg	100	150	200
Water	kg	140	160	180
W/P		1.4	1.1	0.9

Lipari pumice is a relatively pure siliceous volcanic rock with considerable mechanical strength.

Mixture type B

Composition   1,000 litres Latium pumice (2—4 mm fraction)
              300 kg limestone
              15 kg diatomite
              2 kg bentonite

respectively in three proportions:

		B_1	B_2	B_3
Portland cement	kg	80	120	160
Water	kg	88	120	144
W/P		1.1	1.0	0.9

Latium pumice is silicious volcanic rock of tufaceous structure, porous and of low mechanical strength. It is contaminated with clay, obsidian, pyrites and sulphur and must be washed to eliminate these impurities.

Mixtures in the above proportions produce, at 28 days and more, materials that will reproduce the properties of a normal concrete. For this reason we have tried to determine a relation, albeit empirical, connecting the modulus of elasticity and the failure load with the proportions of cement (Fig. 8).

For each mix there is a quantity of cement (q_{max}) beyond which the properties of the materials, for example E_{max} and $\sigma_{uc\,max}$ are not significantly improved. Defining now the non-dimensional relations:

$$V = \frac{E}{E_{max}} \qquad G = \frac{\sigma_{uc}}{\sigma_{uc\,max}} \qquad Q = \frac{q}{q_{max}}$$

we find that the relation that best expresses the variation of the properties as a function of the proportion of cement is:

$$V\left(\frac{1-Q}{\theta \cdot G} + 1\right) = 1$$

where θ is a constant characteristic of the material.

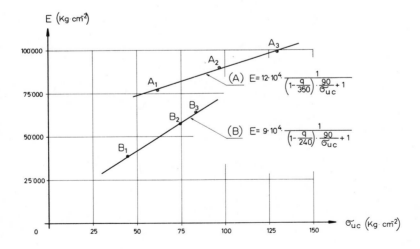

Fig. 8. Relation between elastic modulus and compressive strength as a function of the proportion of cement

Resolving with respect to E and substituting for the product θG the ratio σ_{uc}/σ_0 where $\sigma_0 = \sigma_{uc\ max}/\theta$ we obtain:

$$E = E_{max} \cdot \frac{1}{(1-Q)\dfrac{\sigma_0}{\sigma_{uc}}+1}$$

σ_0 varies between 20 and 90 kg/cm^2 for pumice mixes and from 200 to 300 κg/cm^2 for normal concretes.

It is easy to show that the two mixtures are useful in the whole field we have explored and also beyond, and can be used for ratios of efficiency ζ in the range:

for mixture A: $2.5 < \zeta < 4$

for mixture B: $4 < \zeta < 9$

The usual areas of use for the pumice mixtures for modelling of dams are: mixture A for large scale models and mixture B for small scale models.

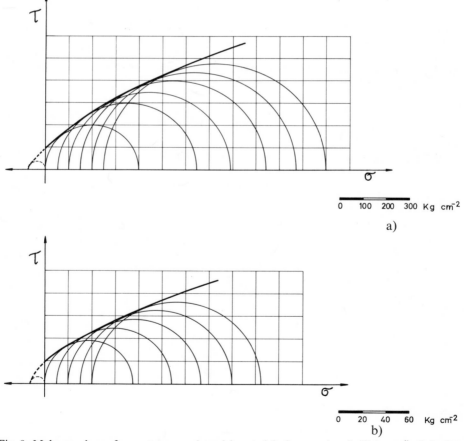

a)

b)

Fig. 9. Mohr envelopes for prototype and model materials for a ratio of efficiency $\zeta = 5,3$: for a) concrete, b) a pumpice mix

These materials, being by nature cemented conglomerates, reproduce faithfully, within their usual range of variability, the properties of normal concretes. Further, tension tests using the Brazilian method show an ultimate tensile strength of the order of 7—9% of the ultimate compressive strength. Poisson's ratio lies between 0.18 and 0.20.

The only significant difference in properties is that the Mohr envelope of the model material is slightly flatter at high normal stress levels, due to crushing of the pumice granules (Fig. 9). However, such states of stress are exceptional for normal structures such as dams; in other cases the error will result in an enhanced factor of safety for the real structure.

On the other hand it is important to consider the shrinkage due to drying out. The strains induced by shrinkage are between 0.5 and 1.0×10^{-3}, that is to say 4 to 8 times greater than for a normal concrete.

For this reason we can resort to the "technique of humid models", covering the model with impermeable varnishes (vinyl and epoxy resins) that impede the evaporation of water [7]. Fig. 10 shows the behaviour of 4 samples of mixture type A treated in this way:
1. rendered impermeable and then kept at 18° C and 80% relative humidity;
2. rendered impermeable and kept as sample 1. for 60 days and then transferred to conditions of 30° C and 50% relative humidity;
3. treated as sample 1. without being made impermeable;
4. treated as sample 2. without being made impermeable.

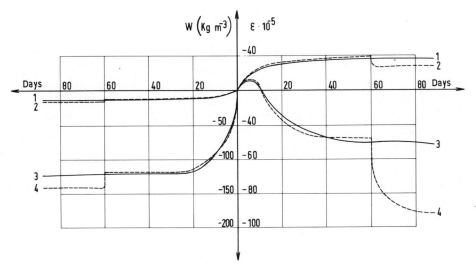

Fig. 10. Shrinkage diagrams for type A mortars

Rendering the material impermeable gives these particular advantages:
homogenous maturing of the interior of the mass;
elimination of shrinkage;
elimination of local hard spots that disturb the rupture process.

Pumice-cement mixtures are suitable for the reproduction of large models for use both within the elastic range and for tests to complete failure in the case of dams for which $20 < \lambda < 150$.

Not all types of extensometers are suitable for use with these materials, in particular the high humidity makes it difficult to satisfactorily glue electrical resistance strain gauges to the model. Nevertheless, the technique of large models makes it possible to employ more sensitive extensometers applied on sufficiently large gauge lengths.

Recently materials have been studied in which the pumice is replaced by a granular aggregate of expanded clay. The principal advantages of this are:

greater homogeneity and consistency of the aggregate;

appreciable reduction in the quantity of water required and of the resulting shrinkage, due to the granules being superficially impermeable.

The lack of fine fractions in the expanded clay necessitates the addition of other aggregates (fine pumice, diatomite, powdered limestone, etc.).

The preliminary results obtained with this material are very encouraging.

2.3.4 Materials for Geomechanical Models

We have seen in chapter 1 that for the reproduction of geomechanical models we must necessarily employ materials with a ratio of efficiency $\zeta \triangleq \lambda$ where λ is of necessity very large, normally greater than 100.

We can produce materials with greatly reduced mechanical properties using mainly pulverulent materials, possibly heavy, bound together with small amounts of binders.

At Ismes we have for many years used mixtures based on litharge (PbO) bound with plaster [5]. The results are again sensitive to the proportion of water and to the setting time in analogy with the mixtures of plaster and diatomite.

As an illustration we can produce a mixture which by varying the proportions is suitable for models of scale $100 < \lambda < 200$ (Table 4). The ratio modulus of deformability/ultimate compressive strength E_r/σ_{uc} (see Chapter 1.4) for such mixtures lies between 700 and 1,000.

Table 4. *Properties of some PbO and Plaster Based Mixes. Composition in % by Weight*

Mix	PbO	Plaster	H_2O	Bentonite	E (kg/cm^2)	σ_{uc} (kg/cm^2)	γ (ton/m^3)	E/σ_{uc}
1	76.0	6.3	16.3	1.4	3,000	3.0	3.65	1000
2	75.0	7.5	16.2	1.3	4,000	5.3	3.61	755
3	74.0	8.7	16.1	1.2	5,500	7.7	3.58	714

We note also the recent work of Barton [1] on a mixture of plaster, Pb_3O_4 (red lead), sand, ballotini and water that satisfactorily reproduces the required mechanical properties for the construction of models reduced in scale $\lambda \triangleq 500$

with the scale ratio ρ approximately equal to 1.3. The properties of the materials that he investigated (Table 5) were in the following range:

$$\sigma_{uc} = 0.53 \text{ to } 2.63 \text{ kg/cm}^2$$
$$E = 210 \text{ to } 1,050 \text{ kg/cm}^2$$

which correspond to prototype values, according to the given scale ratios of:

$$\sigma_{uc} = 350 \text{ to } 1,750 \text{ kg/cm}^2$$
$$E = 140,000 \text{ to } 700,000 \text{ kg/cm}^2$$

Table 5. *Properties of some Mixtures for Geomechanical Models. Composition by Weight*

Mix	Pb_3O_4	Sand/ ballotini	Plaster	H_2O	σ_{uc} (kg/cm^2)	E (kg/cm^2)	γ (ton/m^3)	E/σ_{uc}
A_1	600	1200/0	75	435.0	0.72	252	1.97	350
A_2	600	1200/0	100	442.5	1.25	510	1.96	410
A_3	600	1200/0	125	450.0	2.24	1010	1.98	450
B_1	600	0/1200	75	397.5	1.09	554	1.96	510
B_2	600	0/1200	100	405.0	2.01	1120	1.96	560
B_3	600	0/1200	125	412.5	3.48	1790	1.95	520
C_1	600	600/600	75	416.0	0.60	280	1.95	4.70
C_2	600	600/600	100	423.5	1.41	750	1.94	530
C_3	600	600/600	125	431.0	2.08	1015	1.93	480

The mechanical strength properties of these materials were thoroughly investigated through uniaxial compression tests, tension tests and triaxial tests. We see that the ratio E/σ_{uc} for the materials shown in the table is suitable for a high quality rock matrix, compact and without microfissures. From the publication quoted the method for determining E, defined simply as Young's modulus, is not given.

Finally, at Ismes we have developed new materials that are very suitable for the present purpose. They use:

epoxy resins emulsified in water as binders;

solutions of glycerine and water as diluents and humidifiers;

powdered limestone, barytes or litharge as aggregates;

diatomite and bentonite as additives.

Particular care must be taken when homogenising the material because of the tendency of the resin dispersed in water to separate and coagulate. For this reason, after mixing when it has the consistency of a wet soil, the material is extruded through a die type mincing machine and hence freshly blended.

The particular advantages of these materials are:

rapid hardening through polymerisation of the resin;

large reduction of the evaporation of the water through the presence in solution of the glycerine that greatly reduces the vapour pressure, thus assuring sufficient humidity of the material in equilibrium with its external surroundings, leading to a drastic reduction of shrinkage and resulting internal stresses. This is a most important result for these types of materials.

Table 6. *Properties of some Mixtures for Geomechanical Models. Composition in % by Weight*

Mix	Pumice (pulverized)	Barytes (pulverized)	H_2O	Glycerine (bidistillate)	Epoxy resin	Hardener	E_r (kg/cm²)	σ_{uc} (kg/cm²)	γ (ton/m³)
1	11.80	80.80	5.50	1.24	0.33	0.33	2,500— 3,500	4— 5	2.45
2	11.90	81.05	5.00	1.25	0.40	0.40	4,000— 4,500	6— 8	2.41
3	11.70	81.15	5.00	1.25	0.45	0.45	7,000—10,000	10—12	2.38
4	11.60	81.20	4.95	1.25	0.50	0.50	7,500—11,500	12—13	2.35

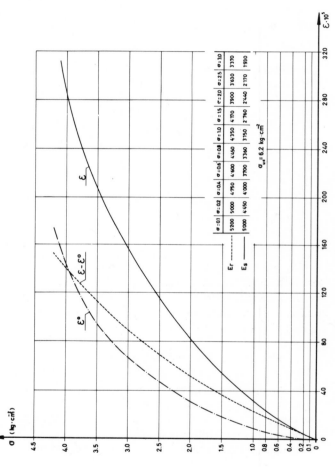

Fig. 11. Behaviour of materials designed for geomechanical models

Such materials have been found particularly suitable for the production of block models [6], [9].

The freshly cast blocks are taken from their moulds and placed in ovens at 60 to 80° C. After 48 hours they attain a condition of good stability, ready for use.

It is important to realise that the properties of rock materials vary between wider limits than those of concrete. As a rule we can fix the ratio E_r/σ_{uc} as being between 300 and 1,000. These limits can be covered in different scale ratios by suitably varying the aggregates and the mix proportions.

Table 6 gives details of several mixes together with their mechanical properties, from which we can obtain values of the ratio E_r/σ_{uc} between 500 and 900.

These materials are normally used for the reproduction of massive rock type models in the scale $100 < \lambda < 200$. For geomechanical model materials, referring to the discussion in Chapter 1.4, the data relative to E_r (modulus of deformability for reversible cycles) have been determined as Ismes for value of σ between 10 and 20% of σ_{uc}.

In this context Fig. 11 shows the behaviour of materials from a mixture of the type 2 as shown in Table 6. From this diagram we can see the wide limits of variability of the secant modulus of total deformability E_s.

We should note however that in geomechanical models, as in nature, the initial cycles for first load are always non-repeatable as distinct from succeeding cycles, when the structure has become stabilized. In general, the repeatability or otherwise of a cycle is governed by the condition that in the interval we do not surpass the previously attained maximum value of load.

From this emerges the complexity of the relative tests and the need for a complete and logical programming of the test cycles together with a continuous awareness of all the deformation processes permanently accumulated in the structure during the preceding cycles, by analogy with that shown digrammatically in Fig. 3.

References

[1] Barton, N. R.: A Low Strength Material for Simulation of the Mechanical Properties of Intact Rock in Rock Mechanics Models. Proceedings of the Second Congress of the I.S.R.M., Beograd, 1970, pp. 3—15.

[2] Cruz Azevedo, M., Esteves Ferreira, M. J.: Construction of Models of Concrete Dams for Elastic Tests. L. N. E. C., Bulletin Nr. 232, Lisbon (1964).

[3] Deere, D. U.: Geologic Considerations. Rock Mechanics in Engineering Practice, Ch. 1 (Stagg-Zienkiewicz, eds.). London: J. Wiley. 1968.

[4] Fialho Lobos, J. F.: The Use of Plastic for Making Structural Models. Rilem, Bulletin, New Series Nr. 8, 65—74, Paris (September 1960).

[5] Fumagalli, E.: Communications sur les matériaux pour modèles statiques de barrages en béton. 5th Congress on Large Dams, Communication C.26, Paris (1955).

[6] Fumagalli, E.: Model Simulation of Rock Mechanics Problems. Rock Mechanics in Engineering Practice, Ch. 11 (Stagg-Zienkiewicz, eds.). London: J. Wiley. 1968.

[7] Fumagalli, E.: Matériaux pour modèles réduits et installations de charge. Ismes, Bulletin Nr. 13 (April 1959).

[8] Fumagalli, E.: Tecnica e materiali per la modellazione delle rocce di fondazione di sbarramenti idraulici. Ismes, Bulletin Nr. 17 (May 1962).

[9] Fumagalli, E.: Modèles géomécaniques des réservoirs artificiels: matériaux, technique d'essais, exemples de reproduction sur modèles. Ismes, Bulletin Nr. 26 (October 1964).

[10] Hobbs, D. W.: The Behavior and Simulation of Sedimentary Rocks. National Cool Board, Mining Research Establishment, Isleworth, Middlesex (1966).

[11] Holdvridge, D. A., Walker, E. G.: The Dehydration of Gypsum and the Rehydration of Plaster. Trans. Brit. Cer. Soc., Vol. 66, 485—509 (1967).

[12] Johnson, R. P.: Strength Tests on Scaled-down Concretes Suitable for Models, with a Note on Mix Design. Magazine of Concrete Research, Vol. 14, Nr. 40. London: Cement and Concrete Association (March 1962).

[13] Lee, J. A. N., Coates, R. C.: The Use of Gypsum Plaster as a Model Material. Civil Engineering and Public Works, Review, London, Vol. 52, Nr. 617, 1261—1263 (November 1957).

[14] Neville, A. M.: A General Relation for Strengths of Concrete Specimens of Different Shapes and Sizes. Journal Americal Concrete Institute, Vol. 63, Nr. 10 (October 1966).

[15] O'Kelly, B. M.: Physical Charges in Setting Gypsum Plaster. Technical Paper 75, ASTM, Bulletin Nr. 237 (April 1959).

[16] Preece, B. W., Sandover, J. A.: Plaster Models and Reinforced Concrete Design. Structural Concrete, Reinforced Concrete Association, London, Vol. 1, Nr. 3, 148—154 (May—June 1962).

[17] Raphael, J. M.: Properties of Plaster-Celite Mixtures for Models. Symposium on Concrete Dam Models, Lisbon (October 1968).

[18] Rauganatham, B. V., Subba Rao, K. S., Hendry, A. W.: Plaster Mortars for Small Scale Tests. ACI Journal, Proceedings, Vol. 64, Nr. 9, 594—601 (September 1967).

[19] Riddel, W. C.: Physical Properties of Calcined Gypsum. Rock Products, Vol. 53, Nr. 5 (May 1950).

[20] Ridge, M. J.: Mechanism of Setting of Gypsum Plaster. Reviews of Pure and Applied Chemistry, Vol. 10, Nr. 4 (1960).

[21] Roll, F.: Materials for Structural Models. ASCE, Vol. 94, Nr. ST6, 1353—1381 (June 1968).

[22] Ross, A. D.: The Effects of Creep on Instability and Indeterminacy by Plastic Models. The Structural Engineering, London (August 1946).

[23] Rowe, R. E., Base, G. D.: Model Analysis and Testing as a Design Tool. Proceedings, Institution of Civil Engineers, London, Vol. 33, 183—199 (1966).

[24] Russell, J. J., Blakey, F. A.: Physical and Mechanical Properties of One-Cast Gypsum Plaster: Plaster AB/2. Australian Journal of Applied Science, Vol. 7 (1956).

[25] Sabnis, G. M., White, R. N.: A Gypsum Mortar for Small-scale Models. ACI Journal, Proceedings, Vol. 64, Nr. 11, 767—774 (November 1967).

[26] Saucier, K. L.: Development of Material for Modeling Rock. Miscellaneous Paper Nr. 6, 934, U.S. Army, Engineer Waterways Experiment Station, Vicksburg, Miss. (October 1967).

[27] Serafim, L., Da Costa, P.: Methods and Materials for the Study of the Weight Stresses in Dams by Means of Models. L. N. E. C., Bulletin Nr. 154, Lisbon (1963).

[28] Simonds, A. M.: Construction of the Plaster and Celite Models of Hoover Dam. Bur. Reclamation Techn. Nr. 306 (1932).

[29] Swaminathan, K. W., Prabhakara, M. K.: Vermiculite for the Preparation of Models for the Analysis of Concrete Structures. Indian Concrete Journal, Bombay, Vol. 38, Nr. 1, 6—10 (January 1964).

[30] White, R. N., Sabnis, G. N.: Size Effects in Gypsum Mortars. Journal of Materials, Vol. 3, Nr. 1, 163—177 (March 1968).

3. Load Application and Techniques of Measurement

3.1 Application of Load — Introduction

We now review the more suitable methods of applying loads to the models.

It must be remembered in this context that the application of loads of the body force type frequently leads, in practice, to problems that are not easily solved and in relation to equation (10) can frequently be a major factor in determining the choice of the fundamental scale.

For example the use uf materials with ratios of efficiency $\zeta = 1$ is frequently dictated by the need to reduce the power requirements of the loading system and the encumberance it causes.

It appears useful therefore to examine at this time the appropriate criteria for choosing suitable methods for practical experiments.

3.2 Gravitational Forces (Self Weight)

The application of gravitational forces to the model is necessary in two cases:

a) when we are investigating the stress distribution due to self weight; in which case we must normally provide a loading and unloading system capable of producing the test load cycles;

b) when in the course of tests to be ultimately taken to failure the self weight represents a significant component of load in relation to the equilibrium of the structure. In this case the self weight can simply be applied as a fixed predetermined and stabilised load.

Here it should be noted that models designed for testing to failure may already, in the course of testing in the elastic region, have been subjected to local plasticity in the region of rigid joints and abrupt changes of section. The absence of self weight can in this case influence both the magnitudes of the global deformations and the local ones in the zones of plasticity.

In the case of slender and simple structure (arches, shells, slabs, frameworks, columns) the self weight forces can if required be applied for simplicity as surface forces.

When we wish to determine the stress distribution in massive structures we can resort to a fictitious system of vertical forces sufficiently distributed through the mass.

In the case in which the load is required only as a stabilising component two solutions are available:

i) appropriately modify the density of the model material with reference to equation (10) by the use of heavy aggregates. This solution is, in practice, only of limited use. We have seen in fact how in the case of microconcrete, we can operate within a limit $\rho \max \geq {}^1/_2$. For materials for geomechanical models for which $\zeta \triangleq \lambda$ it is often necessary to increase the weight of the material with heavy powders (litharge, red lead, barytes) to satisfy the relation $\rho \triangleq 1$;

ii) resort to a fictitious system of discret vertical forces, as discussed in the following section.

Other ingenious methods, suitable within limits and for particular cases, will be illustrated below.

3.2.1 Fictitious Gravitational Forces

The general method consists of artificially distributing in the interior of the model a sufficient and rational system of vertical forces, causing the minimum of encumberance, in a way that disturbs as little as possible the static behaviour of the model.

In the technique in use at Ismes [6], the total volume of the model is subdivided into elementary volumes, roughly parallelepipeds (ideally cubes), estimating their centres of gravity and the intensity of the equivalent force to be applied.

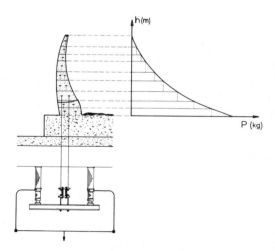

Fig. 12. Schematic representation of the loading system for self weight for a model of an arch dam

A rod of high tensile steel is passed vertically through a plastic tube and fixed by cold riveting to a metal disc which itself is fixed to a larger diameter plastic disc placed at the centre of gravity of each elementary volume. Placing the steel rod in the plastic tube effectively isolates it from the surrounding model. The plastic disc has a modulus almost equal to that of the model material and a sufficiently large diameter to avoid stress concentrations and local failure (Fig. 12).

An installation of this type causes little encumbrance, the volume in the case of an arch dam rarely exceeds 1% of the total volume of the model.

The choice of a suitable means of applying the loads depends on the type of tests being planned.

If the application of this load is required simply to stabilise the structure against the action of other load components (wind loading, hydrostatic loading, etc.) a fixed system using spring dynamometers, duly calibrated, is generally sufficient.

The dynamometers each apply load through a single loading rod or, using a balancing system, through two or more loading rods. Their reactions are taken by an anchor plate fixed in reinforced concrete or by a steel framework placed under the foundation of the structure supporting or containing the model. Application and adjustment of their load is effected through a screw with a manually adjusted nut and lock nut.

If we wish to measure the stress distribution in models above a certain size, it is not possible in practice to provide a manual loading system that can be operated in a reasonable period of time. Since for tests in the elastic region it is not advisable to prolong the duration of loading cycles beyond a reasonable time (15 to 20 minutes), then in such a case the loading and unloading operation may be achieved with the aid of a battery of jacks acting on a sufficiently rigid movable steel anchor plate.

The vertical displacement of the anchor plate produces an equal lengthening of all the interposed dynamometers, so that the force applied to each of the loading rods is proportional to the spring constant of the dynamometer, K (kg/cm).

For applications of this type, spring dynamometers of the type referred to have the inconvenience of undergoing too limited an elongation (2 to 4 cm). Thus small errors resulting from the deformability and imperfect levelling of the anchor plate are not negligible. It is more practical to use rubber rings produced for this purpose by the Central Research Laboratory of the "Società Pirelli".

The rubber used for these rings is characterised by a very high limit of elasticity in extension, having a load-deformation relation that, while not being perfectly linear, shows very little hysteresis or residual anelasticity. The properties of these rings cannot be compared for precision with those of a properly stabilised and calibrated steel spring but against this they tolerate extensions up to 15 to 20 cm without yielding.

Such high elongations lead to a proportionate reduction of the fixed errors resulting from imperfect levelling or deformation of the movable anchor plate so that the total error is of the order of 2—3%, certainly less than that of a spring dynamometer system.

The application of load through rings, besides satisfying the criteria for extremely practical assembly and use with the minimum of encumbrance, can be further improved by placing groups of rings at different heights.

Fig. 13 shows an example of the application of self weight loading to a model of an arch dam.

3*

Fig. 13. Example of a self weight load application system for an arch dam model

As an illustration of a practical application we can take that of the model of the arch dam of "Almendra" (Spain) reproduced at a scale 1:100 (480 m chord, 222 m high). 560 load rods were used, each one of which was applied to a volume varying from 2.5 dm³ at positions corresponding to the region of the crest to 7.5 dm³ in the region of the plug.

In the case of arch dams the stress distribution due to selfweight in the real structure is appreciably influenced by the construction method used. In particular only a successive programme of injection of the vertical joints renders the work monolithic in respect of the succeeding masses and stages.

In the model it is still possible to apply the self weight load in two or three increments, resulting from two or three independent settings of the self weight loads following the injection of the joints of the model dam in two or three successive stages, making separate measurements as the weight of the succeeding stage, with joints open, distributes itself in the structure below that has had its joints closed in accordance with the executive technique and plan of the contractor.

3.2.2 Self Weight by Centrifuging

This method used by, among others, the Vadeneev Hydrotechnical Institute of Leningrad [16] consists of rotating the model in a horizontal plane with constant angular velocity.

The method is interesting because of the high effective weights attainable in the acceleration centrifuge, but demands, for accuracy, that the radial dimensions of the model be small compared with the radius of gyration of the centrifuge (normally not greater than $^1/_{10}$).

A particular difficulty of this method is the problem of providing moving contacts (slip rings) for transferring the measurements to the position for reading and external recording.

The method has the advantage of making possible experiments on geomechanical and geotechnical models with a ratio of efficiency $\zeta = 1$.

The method has been shown to be useful in the investigation of models concerned with the stability of slopes in loose ground (landslips).

3.2.3 Self Weight by Inversion

This method depends on determining the change in stress distribution within the model when it is turned upside down; this being equivalent to applying to the model a system of forces equivalent to twice its own self weight [19].

Working on models at much reduced scale it is necessary to use materials of high density and deformability. This places limits on the application of the method for the reasons given in paragraph 3.2.

In some practical applications polyester resins filled with barytes and litharge have been used as model materials.

3.2.4 Distribution of the Self Weight for Successive Stages of Construction

This method, developed by Raphael, is used in practice by the L.N.E.C. of Lisbon [19]. It enables an evaluation to be made of the distribution of the self weight load during the execution of the construction.

It is a particularly suitable method for shell dams in that it permits the determination of the final stress distribution as the sum of all the stress states built up in the structure during the intermediate stages of construction.

Since in practice it is impossible to reproduce all the stages of construction, the test is conducted as follows.

On to the structure, built up to an arbitrary heigh h (Fig. 14), is applied a uniformly distributed load of intensity p, sufficiently large to produce easily measurable deformations.

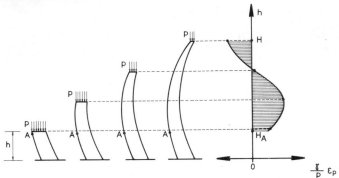

Fig. 14. Outline of the method of construction by layers applied to a cantilever of unit width

ε_p is the strain at an arbitrary point A due to this load.

The strain corresponding to a layer of material of height dh and specific weight γ is given by

$$\frac{\varepsilon_p}{p} \, \gamma \, dh$$

On completion of construction the total strain at A becomes:

$$\varepsilon_A = \int_{H_A}^{H} \frac{\varepsilon_p \, \gamma}{p} \, dh = \frac{\gamma}{p} \int_{H_A}^{H} \varepsilon_p \, dh \, .$$

In practice it is sufficient to determine experimentally a finite number of values $\frac{\gamma}{p} \cdot \varepsilon_p$ so as to enable the curve of the derivative to be traced as shown in Fig. 14, and then to proceed to graphical integration.

In the same way we can determine the displacements.

Since it is easier to cut down a model rather than to build it up in layers, the method is applied in reserve. For example, in the case of dams, after having carried out the hydrostatic load tests we proceed to remove successive horizontal layers from the body of the dam (Fig. 15).

Fig. 15. Example of application of the layer method on a multiple arch dam

3.3 Surface Forces

The intensity of the specific force (kg/cm²) to be applied to the model does not depend on the geometric scale of reproduction λ but only on the ratio of efficiency ζ of the material [11].

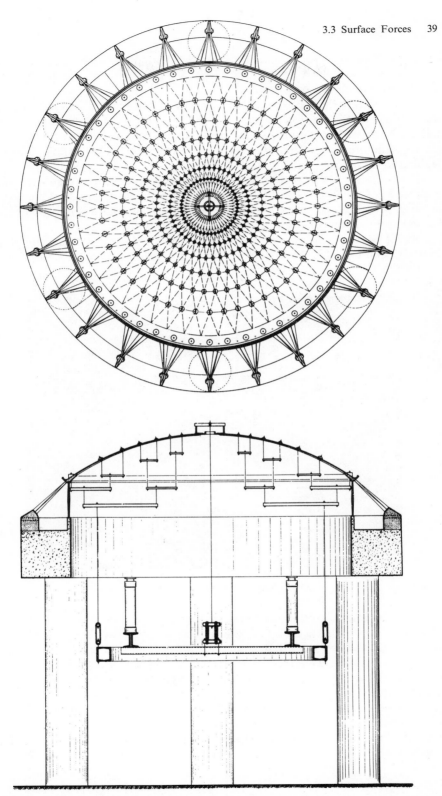

Fig. 16. Scheme of application of surface forces by means of ring dynamometers for a shell roof

For surfaces of roofs, floors etc. an elementary method, but frequently the least rational, is that of applying the local directly to the surface by means of ballast: brick, concrete blocks, bags of sand.

Often we have the problem that covering the surface makes it difficult to make measurements. Besides for highly deformable structures, particularly in tests to failure, it can happen that the ballast accumulated on the surface acts as a continuous structure that effectively increases the stiffness of the whole assembly, leading to significant errors in the results.

Better one should consider when possible suspending the load below the model by means of loading rods. The surface is divided into a sufficient number of points, to each of which we apply the corresponding load by means of hooks passed through the model and then bolted in position.

At the point of application of the load to the surface it is necessary to place a load distribution plate to avoid rupture in the materials by the concentration of stress.

Fig. 17. Example of the application of surface forces by means of ring dynamometers

In the simplest case the load may be applied to the loading rods by means of weights. To facilitate cycles of loading and unloading we can use a movable platform operated by jacks that can be raised to support the load of the weights.

For graduating the load we can resort to a placing a series of weights at several positions in a vertical chain so that the weights can be progressively brought into action by lowering the movable platform in successive stages.

For small models we can use as weights containers filled with lead shot.

At Ismes we have made great use of load application using rings in exactly the same manner as that already described in the case of the application of self weight (Fig. 16 and 17). Installations of this type are much more rational and cause less encumberance and in addition they allow the application of loads in a gradually increasing and continuous manner.

Less frequently we apply the load directly to the surface by means of hydraulic jacks and distribution plates.

3.4 Hydrostatic Pressure

The hydrostatic pressure must be treated as a local pressure to be reproduced in the scale ζ, the same as for any specific surface force. We note that as a body force the relative specific weight is governed by equation (10).

The application of the hydrostatic thrust is of major importance in modelling dams, an area among the most developed and advanced in the technique of modelling.

3.4.1 Method of Liquids of Different Densities

Application of hydrostatic pressure by means of a liquid certainly provides a method which is both exact and easily applied; however the method is not generally applicable in ordinary practice for several reasons.

It is necessary in the first place to realise that structures of large size, such as dams, require a very large reduction in the scale ratio λ and, excluding those cases in which for other reasons it is necessary according to equation (10) to reduce the scale of ζ in more or less the same ratio (as for example for geomechanical models), the densities of the available liquids are insufficient.

Further difficulty results from the practical impossibility of continuously increasing the density of the loading liquid to increase the loads in the course of the experiments.

Nevertheless they are widely used for elastic models in materials of high deformability or for geomechanical models.

The liquid is made to flow into and out of one or more rubber bags placed between the surface under load (upstream face of a dam, impermeable barrier, etc.) and a rigid surface shaped to match that of the model and positioned 1 to 2 cm from it.

Suitable and economically available liquids that are worth mentioning are:

	Specific gravity of the pure liquid or saturated solution at 20° C.	Solvent
Potassium Carbonate (K_2CO_3)	1.5	water
Zinc Chloride ($ZnCl_2$)	1.7	water
Zinc Chloride in acid solution	1.9	water
Mercury potassium iodide ($MgI_2 \cdot 2\,KI$)	3.1	water
Mercury	13.5	—

All the above liquids have been used in practice.

For short term use and for suitable quantities of liquid suspensions can be economically employed. For example a saturated acid solution of zinc chloride with the addition of bentonite and powdered barytes can provide a sufficiently stable and homogeneous suspension for tests of normal duration giving a relative density of 2.7 to 2.8.

The maximum density is governed by the allowable limit of fluidity for hydraulic transport.

For greater densities of the order of 3 it is necessary to carefully decant the suspension.

The biggest problems are associated with preservation since in time, in the absence of agitation, the solid phase settles slowly catalysing crystalization of the zinc chloride.

3.4.2 Method of Bags of Liquid with Differential Hydraulic Head

This technique is useful when one requires an inexpensive loading system that is easy to regulate.

Basically this method involves placing a series of liquid filled bags in horizontal strips on the model surface and providing a mechanism for simultaneously regulating the hydraulic head in all the bags [6].

For this purpose each bag is connected to a small reservoir fixed at a precise point on an arm hinged at the point 0 at the level of maximum water level behind the dam and initially placed in the fully lowered resting position (Fig. 18). If we now lift the arm, by rotating about the hinge at 0, until it is horizontal we produce

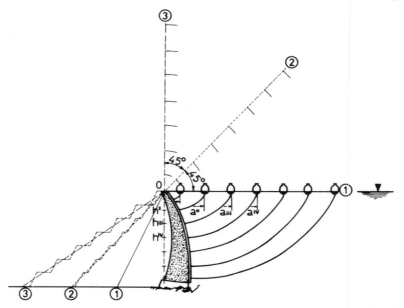

Fig. 18. Scheme of application of hydrostatic load using bags of liquid with differential hydraulic heads

a perfectly triangular thrust diagram whose slope is determined by the density of the liquid used.

The reservoirs are positioned along the movable arm at distances $a', a'' \ldots$ from 0 proportional to the depth $h', h'' \ldots$ of the mean level of the corresponding bag, according to a constant ratio n, that is to say:

$$\frac{a'}{h'} = \frac{a''}{h''} = \frac{a'''}{h'''} = n$$

Now raising the arm until it is inclined at an angle α to the horizontal the liquid produces a thrust as if its density had been increased such that:

$$\gamma' = \gamma \, (1 + n \sin \alpha)$$

We note that the resulting pressure distribution is not perfectly triangular, due to the addition of a constant pressure head to each bag. A closer approximation of the stepped loading diagram to the true one can be obtained by increasing the number of bags.

If we wish to arrange for some free surface for measurement purposes we can separate all the bags by a distance $\Delta h'$ leaving the bags a height $\Delta h''$ so that:

$$K = \frac{\Delta h''}{\Delta h' + \Delta h''}$$

and the apparent density of the liquid becomes:

$$\gamma' = K\gamma \, (1 + \sin \alpha)$$

3.4.3 Method of Hydraulic Jacks

This method finds more general application at Ismes than all the other methods that have been discussed, above all for practical reasons [6].

It allows one to:

easily regulate the hydraulic pressure in the jacks, and consequently the apparent density of the thrusting liquid, in a continuous manner over a wide range;

provide free surface on the upstream face for the application of measuring equipment, yet ensure the complete application of the resultant thrust to the model;

apply to the model an initial load, to absorb the load losses due to initial friction in the hydraulic jacks and the small deformations of initial adjustment.

As an illustration we look at the procedure for setting up a hydrostatic load for an arch dam (Fig. 19). In order to take account in the structure of the releases due to the radial joints we advise that the surface of each individual block be considered separately.

We consider the development of the face in horizontal arches. It is always possible to divide the surface into separate levels.

$$h' - h'' - h''' \ldots h^{m-1} - h^m$$

and to inscribe on the defined surfaces a set of trapeziums or triangles.

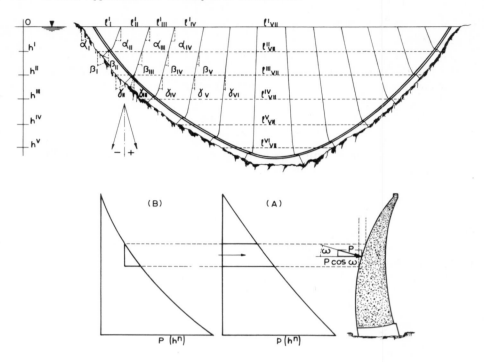

Fig. 19. Procedure for designing the setting up of hydrostatic loads for an arch dam model: (A) diagram of thrust for strips of unit height, (B) diagram of total thrust

The small associated errors are neutralised by compensation on the surfaces of adjacent blocks.

It follows, on all strips of unit height for any block and for the zone included between $(o$ and $h')-(h'$ and $h'')-(h''$ and $h''')$ the actions of horizontal thrust are given by the following expressions.

$$p(h')=\gamma h \left\{l'_n+h(\tan \alpha_n-\tan \alpha_{n+1})\right\}$$
$$p(h'')=\gamma h \left\{l''_n+(h-h')(\tan \beta_n-\tan \beta_{n+1})\right\}$$

where:

$$l''_n=l'_n+h'(\tan \alpha_n-\tan \alpha_{n+1}) \qquad (a)$$
$$p(h''')=\gamma h \left\{l'''_n+(h-h'')(\tan \gamma_n-\tan \gamma_{n+1})\right\}$$

where:

$$l'''_n=l''_n+(h''-h')(\tan \beta_n-\tan \beta_{n+1})$$

The whole load and always the horizontal component alone can be expressed as.

$$P\,(h') = \int_0^h p\,(h')\,dh$$

$$P\,(h'') = \int_0^{h'} p\,(h')\,dh + \int_0^h p\,(h'')\,dh$$

$$P\,(h''') = \int_0^{h'} p\,(h')\,dh + \int_{h'}^{h''} p\,(h'')\,dh + \int_{h''}^h p\,(h''')\,dh \qquad\qquad\qquad (b)$$

If we wish to evaluate the height Δh corresponding to each hydraulic jack, it is useful to draw two diagrams. The one concerning the intensity of load for strips of unit height by the application of expression (a) (diagram A of Fig. 19), the other concerning the horizontal component of the full load as a function of h by the application of the expression (b) (diagram B of Fig. 19).

The first diagram allows us to evaluate Δh corresponding to each jack, taking account of the fact that the first diagram gives, to a satisfactory approximation, the value of

$$P \cos \omega$$

and that the second gives the lines of application of $P \cos \omega$ and that of the centre of thrust.

Each jack applies load to one or two load distribution plates that cover only 60—70% of the surface of the face and are connected to the latter through a sheet of highly deformable material (cork).

To reduce friction to a minimum between the distribution plate and the loaded surface we recommend the introduction of sheets of silver paper coated with molybdenum disulphide.

We ourselves arrange that the jack sizes, in terms of thrust, are matched to the thrusts required on the appropriate surfaces so that all the jacks can be energised to a high pressure from a single pressure source through a suitable network of pipes.

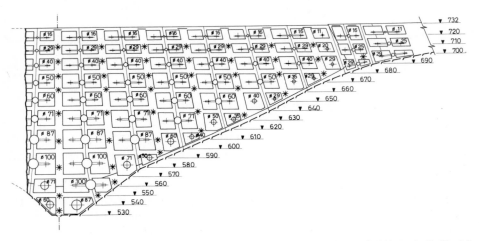

Fig. 20. Scheme of application of hydrostatic load for the arch dam of "Almendra" (Spain)

The oil pressure is normally adjustable up to 600 atm. by means of a sensitively controllable pump.

In Fig. 20 there is shown, as an illustration, the arrangement that was used for applying the hydrostatic load to the model of the arch dam "Almendra" (Spain).

Fig. 21 shows a general view of the same model complete with hydrostatic loading system.

Fig. 21. General view of the model of the "Almendra" arch dam with the hydrostatic load system

3.5   Tests with Combined Resultant Loads

It is worth mentioning a technique devised by the L. N. E. C. of Lisbon [15] for carrying out tests to failure with resultant proportional increases of self weight and hydrostatic thrust.

The structure of the cantilever is divided into blocks to each of which belongs a weight force at the centre of gravity and a superficial hydrostatic thrust that combine to produce an inclined resultant that is applied by means of a jack pushing against the surface of the mass through a suitable distribution system (Fig. 22 and 23).

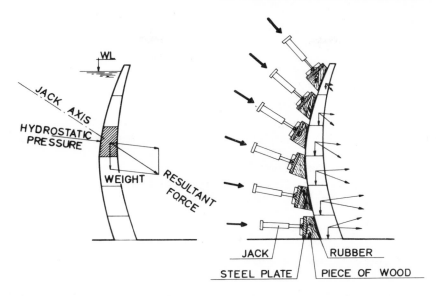

Fig. 22. Outline of the system of loading with hydraulic jacks for the simultaneous application of hydrostatic pressure and self weight for arch dam models

In the case of large deformations (experiments to failure) the shear stresses due to friction between the model surface and the load distribution pad necessary to prevent slipping of the distribution pad can induce some errors due to the additional constrains.

Fig. 23. Example of application of combined resultant loads on a dam model

3.6   Concentrated Forces

Concentrated forces must be applied according to the relationship

$$\Psi = \zeta \, \lambda^2$$

No matter how concentrated, a force must always be distributed over an area; for this reason it is always advisable to ascertain the contact pressure and the need to resort to local frames or distribution systems.

These loads are normally applied by hydraulic jacks acting directly at the point of application of the force. When possible, it is preferable to apply these loads with loading rods since these latter constrain the structure less.

3.7   Instruments and Measurements

It is not possible in the present work to fully examine the very large range of measurement techniques and applications. This is not the purpose of the present publication.

We will limit ourselves therefore to some relevant comments in specific fields of use.

The fundamental measuring instrument is the extensometer. The normal sensitivity that is required is of the order

$$\varepsilon = 1 \text{ to } 5 \times 10^{-6}.$$

The accuracy of the extensometer, considering the relative importance that this term acquires in the field of models, must be approximately equal to its sensitivity.

It must be noted that the total error incurred in the measurement of a quantity is always the sum of accidental and systematic errors.

Among the accidental errors we note in particular:

local variations of the material properties compared with the values found from the test samples;

errors of judgment of the operator;

imperfections in mounting the instruments;

the effect of temperature variations etc.

Among the systematic errors we can record those due to:

imprecise calibration of the instruments;

initial inertia of the instruments;

imprecise determination on the test sample of the mechanical properties of the material (modulus of elasticity, Poisson's ratio, etc.);

imprecise application of loads.

It is easier to guard against accidental errors. In particular this can be done by taking readings in four directions, that is to say as two pairs of orthogonal readings. It is then possible to check the equality of the scalar invariant for the two pairs of readings[2].

For the same purpose it is normally useful to examine also the coherence of the pattern of the stress distribution, coherence being most easily monitored through examination of the principal stresses.

---

2 To check a four arm rosette arranged in the following pattern:

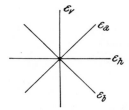

we proceed to check that the invariant $(\varepsilon_a+\varepsilon_b)=(\varepsilon_v+\varepsilon_h)$; if these are not equal we normally adjust the measurements until we obtain a satisfactory approximation.

The residual error $\Delta=(\varepsilon_v+\varepsilon_h)-(\varepsilon_a+\varepsilon_b)$ can in each case be distributed in compensation in equal parts $\left(\dfrac{\Delta}{2}\right)$ between the two pairs of readings according to a criteria of mean weighting for the individual readings.

This adjustment criterion is realised by applying the following formulae:

$$\varepsilon_{v\ corr.}=\varepsilon_v-|\varepsilon_v|\cdot\frac{(\varepsilon_v+\varepsilon_h)-(\varepsilon_a+\varepsilon_b)}{2\,(|\varepsilon_v|+|\varepsilon_h|)}$$

$$\varepsilon_{h\ corr.}=\varepsilon_h-|\varepsilon_h|\cdot\frac{(\varepsilon_v+\varepsilon_h)-(\varepsilon_a+\varepsilon_b)}{2\,(|\varepsilon_v|+|\varepsilon_h|)}$$

$$\varepsilon_{a\ corr.}=\varepsilon_a+|\varepsilon_a|\cdot\frac{(\varepsilon_v+\varepsilon_h)-(\varepsilon_a+\varepsilon_b)}{2\,(|\varepsilon_a|+|\varepsilon_b|)}$$

$$\varepsilon_{b\ corr.}=\varepsilon_b+|\varepsilon_b|\cdot\frac{(\varepsilon_v+\varepsilon_h)-(\varepsilon_a+\varepsilon_b)}{2\,(|\varepsilon_a|+|\varepsilon_b|)}$$

The corrected ε give the following values for σ

$$\sigma_v=\frac{E}{1-v^2}\,(\varepsilon_v+v\,\varepsilon_h)$$

$$\sigma_h=\frac{E}{1-v^2}\,(\varepsilon_h+v\,\varepsilon_v)$$

$$\sigma_a=\frac{E}{1-v^2}\,(\varepsilon_a+v\,\varepsilon_b)$$

$$\sigma_b=\frac{E}{1-v^2}\,(\varepsilon_b+v\,\varepsilon_a)$$

From $\sigma_v,\sigma_h,\sigma_a$, we can calculate the principal stresses:

$$\sigma_{I.II}=\tfrac{1}{2}\left\{(\sigma_v+\sigma_h)\pm\sqrt{(\sigma_v-\sigma_h)^2+[2\sigma_a-(\sigma_h+\sigma_v)]^2}\right\}$$

and the angle ϕ between σ_{max} and σ_v is given by:

$$\tan\phi=\frac{2\sigma_a-(\sigma_v+\sigma_h)}{(\sigma_v-\sigma_h)+\sqrt{(\sigma_v-\sigma_h)^2+[2\sigma_a-(\sigma_h+\sigma_v)]^2}}$$

4 Fumagalli, Statical Models

It is more difficult to determine the existance and magnitude of the systematic errors since they can influence proportionately the whole pattern of the results. It is therefore advisable, when possible, to verify the equilibrium between the applied loads and reactions at the boundary acting on the internal structure or a suitably isolated part of this.

For the principles of operation of extensometers (strain gauges) the reader is referred to the bibliography and to the illustrated reports provided by the manufacturers.

We limit ourselves here to some considerations on the advantages, inconveniences, methods and areas of application of these instruments.

The extensometers normally used are:

a) mechanically amplified direct reading extensometers, available for single readings and rosettes, provided that there is easy access for reading. The techniques and accuracy of manufacture of the extensometers produced by the more reputable firms result in instruments with extremely small initial inertia and high accuracy so that they now find wide areas of application for limited numbers of readings where there is easy access, above all because of their simplicity of use (Fig. 24a);

b) electro-acoustic or vibrating wire strain gauges reading out to a remote centralised terminal. They have the indisputable advantage that their readings,

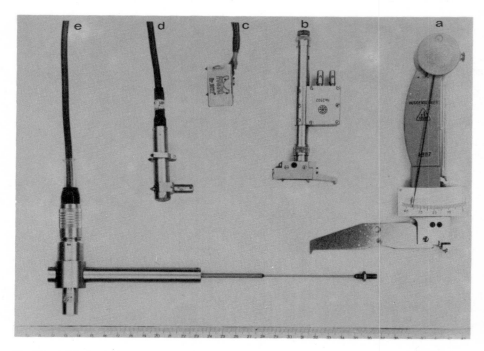

Fig. 24. a) Mechanical extensometer (Huggenberger), b) electro-acoustic extensometers (Galileo). c) electro-optical extensometer (Maihak), d) inductive extensometer (Hottinger), e) inductive displacement transducer (Hottinger)

resulting from the resonant frequency of a stretched vibrating wire, are not subject to errors due to electrical losses resulting from inadequate insulation or poor contacts. They are, therefore, suitable for use in regions where access is difficult and in unfavourable ambient conditions. Readings are obtained by arranging for a variable control frequency in the reading terminal to vibrate in unison with the strain gauge (by elimination of beats). They produce measurements of great precision, certainly not inferior to those of mechanical extensomers (Fig. 24b);

c) electro optical strain gauges; these are derivatives of electro acoustic ones. The principle of operation is the same except that the reading in this case is made optically by observing the Lissajous figures produced on an oscilloscope. In this way even greater sensitivity is obtained but the resulting instrument is somewhat more delicate (Fig. 24c);

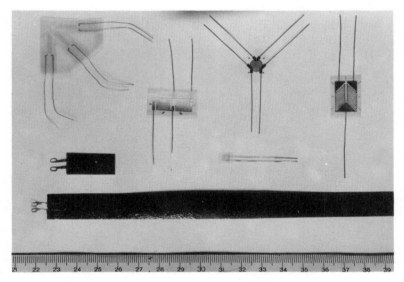

Fig. 25. Electrical resistance strain gauges

d) electrical resistance strain gauges (Fig. 25). These have the indisputable advantage of compactness and thus cause minimal obstruction. They are produced in a wide range of gauge lengths from 1 to 2 mm up to 300 mm and greater.

The smaller gauges are indicated for experiments on small models and for measurement of small details.

Such strain gauges have another important use for automatic digital readout and continuous recording.

Frequently they are used as the sensitive element in more complex instruments (flexible beams and membranes).

They are usually regarded as expendable devices.

They are also indicated for dynamic and moving models.

Among their disadvantages we note:

their sensitivity to humidity;

errors due to resistance losses in connecting cables when measuring at considerable distances;

relatively limited sensitivity, of the order of $\varepsilon = 5 \times 10^{-6}$;

application by glueing requires a perfectly dry surface for mounting[3].

---

3 For the complete determination of the state of stress at a point we normally require to measure strains in three different directions. In this case however it is not possible to detect errors by comparison of scalar invariants together with the appropriate compensation.

The appropriate formulae for deducing the principal stresses are as follows [12]:

Rectangular rosettes

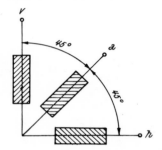

Principal stresses

$$\sigma_{I,II} = \frac{E}{2} \left\{ \frac{\varepsilon_h + \varepsilon_v}{(1-v)} \pm \frac{1}{(1+v)} \sqrt{(\varepsilon_h - \varepsilon_v)^2 + [2\varepsilon_a - (\varepsilon_h + \varepsilon_v)]^2} \right\}$$

$$\tan 2\phi = \frac{2\varepsilon_a - (\varepsilon_h + \varepsilon_v)}{\varepsilon_h - \varepsilon_v}$$

ϕ = angle between the direction $\sigma_{\max}$ and the horizontal axis.

Delta rosettes

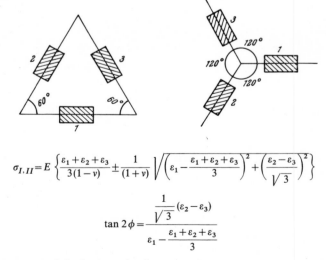

$$\sigma_{I,II} = E \left\{ \frac{\varepsilon_1 + \varepsilon_2 + \varepsilon_3}{3(1-v)} \pm \frac{1}{(1+v)} \sqrt{\left(\varepsilon_1 - \frac{\varepsilon_1 + \varepsilon_2 + \varepsilon_3}{3}\right)^2 + \left(\frac{\varepsilon_2 - \varepsilon_3}{\sqrt{3}}\right)^2} \right\}$$

$$\tan 2\phi = \frac{\frac{1}{\sqrt{3}}(\varepsilon_2 - \varepsilon_3)}{\varepsilon_1 - \frac{\varepsilon_1 + \varepsilon_2 + \varepsilon_3}{3}}$$

ϕ = angle between $\sigma_{\max}$ and the horizontal axis.

e) electrical inductance strain gauges; these instruments have recently found an important field of application since they are more suitable than resistance strain gauges for use with automatic recording systems.

They have a high sensitivity ($\varepsilon = 1.10^{-6}$).

Highly deformable, they are capable of measuring, with the simple instrument, quantities greater than a millimeter.

Different mountings can be employed to enable the instrument to be used to measure either total displacement (Fig. 24e), or relative displacement (strain gauging) (Fig. 24d) used either with digital readout, or with continuous recording.

A modern system fully equipped for digital reading can provide, in quasi automatic form through an electronic data logger, the strains, stresses and when required the principal stresses at the point in question (Fig. 26).

Having set up the system, application to a large model allows one to completely programme the tests with rapid readings covering a large number of points in a relatively short time and carried out at prearranged fixed intervals.

Fig. 26. Data logging equipment for electrical transducers

3.7.1 Demountable Extensometers

When one does not possess an adequate system of instruments, an investigation of the strain distribution in a model (provided that it is sufficiently large) can be achieved more economically by using demountable extensometers.

A well trained operator using for example a demountable Demec gauge (shown in Fig. 27b) can measure deformations to an accuracy of ± 0.2 microns.

Such accuracy on a gauge length of 50—100 mm is entirely satisfactory and not greatly inferior to that normally obtained from extensometers with fixed installations.

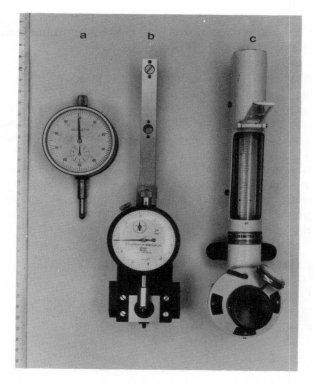

Fig. 27. a) Dial gauge, b) demountable (DEMEC) strain gauge, c) clinometer

Demountable strain gauges are particularly useful for a series of measurements along a line.

The base line can be set out as a single or double chain; in this latter case the measurements of one chain are displaced, with respect to the other placed in parallel, by half a gauge length.

If finally the sum of the displacements is checked with displacement measurements at the extreme ends of the lines it is possible to obtain, with agreement of the measurements, a complete and detailed investigation, frequently of great precision.

3.7.2 Displacement and Clinometer Measurements

Dial gauges for displacement or global deformation measurement are normally available with a sensitivity of $^1/_{100}$ mm (Fig. 27a). For a greater sensitivity we can resort to instruments with a sensitivity of $^1/_{1000}$ mm. In the latter case the mechanical workmanship of the instrument must be of the highest precision, otherwise the errors resulting from stiff gearing can be greater than those of the ordinary gauges.

During experiments to failure the measurements of total deformation are normally made with displacement transducers that use as their sensitive element electrical resistant strain gauges, or, with better results, Hottinger type inductance gauges (Fig. 24e).

In rare cases we resort to the use of clinometers for showing rotations of the structure, particularly with reference to the foundations.

A clinometer suitable for this purpose is the Huggenberger (shown in Fig. 27c) that can measure rotations to an accuracy of 1 second (sexagesimal).

3.8   Brittle Lacquers

It has been noted that small cracks are produced in brittle lacquers (varnish) by tensile stresses that are within the normal working range of the metallic materials on which they are applied.

Cracking of the lacquers shows visibly the development of the isotatics normal to the family of isotatics of maximum tension, during tests in the elastic range.

These lacquers are suitable for use in experiments on steel structures, materials for which the ratio

$$250 < \frac{E}{\sigma_{ut}} < 500$$

more so than for concrete material that are much more fragile for which

$$7000 < \frac{E}{\sigma_{ut}} < 10,000$$

In this latter case in fact rupture of the concrete occurs at approximately the same strain levels as that of the lacquer. A particular application is however in use at the L.N.E.C., Lisbon [18], where the lacquer is applied to the model of a dam while it is subject to hydrostatic loading. After the lacquer has hardened the model is unloaded.

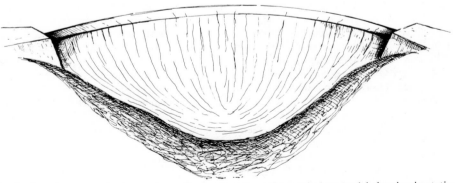

Fig. 28. Brittle lacquer crack pattern on downstream face of an arch dam model showing isostatics normal to the direction of maximum compressive stress

The cracking of the lacquer shows in this case the expansion through decompression of the model material (plaster-diatomite) for which

$$350 < \frac{E}{\sigma_{uc}} < 500$$

To enhance the effect, the lacquer under tension can be cooled with evaporating solid carbon dioxide.

Fig. 28 shows schematically the experimental crack pattern on the downstream face of a model of an arch dam that indicates the development of the isostatics normal to the family of isostatics for maximum compression.

References

[1] Balas, J.: Some Applications of Experimental Analysis of Models and Structures. Experimental Mechanics, Vol. 7, Nr. 3, 127—129 (March 1967).

[2] Benito, C.: Scale Model Testing as an Efficient Aid to the Designer. Rilem, Intern. Colloq. of Mod. of Structures, Madrid (1959).

[3] Biggs, J. M., Hansen, R. J.: Model Techniques Used in Structural Engineering Research at M.I.T. Rilem, Intern. Colloq. of Mod. of Structures, Madrid (1959).

[4] Carpenter, J. E., Magura, D. D., Hanson, N. W.: Structural Model Testing-techniques for Models of Plastic. Journal, PCA Research and Development Laboratories, Vol. 6, Nr. 2, 26—47 (May 1964).

[5] Ellis, G., Stern, F. B., Boranowski, S. J.: Brittle Coatings (Chapter V of SESA Manual of Experimental Stress Analysis). Experimental Mechanics, Vol. 6, Nr. 10 (October 1966).

[6] Fumagalli, E.: Matériaux pour modèles réduits et installations de charge. Ismes, Bulletin Nr. 13 (April 1959).

[7] Hoek, E.: The Design of a Centrifuge for the Simulation of Gravitational Forces Fields in Mine Models. Journal of the South African Institute of Mining and Metallurgy, Vol. 65, Nr. 9, 455—487 (April 1965).

[8] Hossdorf, Heinz: Cable et dispositif tendeur pour précontraindre les modèles réduits en mortiers. Rilem, Intern. Colloq. of Mod. of Structures, Madrid (1959).

[9] Litle, W. A., Hansen, R. J.: The Use of Models in Structural Design. Journal, Boston Society of Civil Engineers, Vol. 50, Nr. 2, 59—64 (April 1963).

[10] Nizery, A., Remenieras, G., Beaujoint, N.: Etude sur modèle réduit des contraintes dans les barrages. Extrait de: Annales des Ponts et Chaussées (July-October 1953).

[11] Oberti, G.: Large Scale Model Testing of Structures outside the Elastic Limit. Rilem, Intern. Colloq. of Mod. of Structures, Madrid (1959).

[12] Perry, C. C., Lissuer, H. R.: The Strain Gage Primer. New York: McGraw-Hill. 1962.

[13] Rocha, M.: Structural Model Techniques. Some Recent Developments. L.N.E.C., Memoria Nr. 264, Lisbon (1965).

[14] Rocha, M., Borges, J. F.: Photographic Method for Model Analysis of Structures. Publication Nr. 18, L.N.E.C., Lisbon (1951).

[15] Rocha, M., Serafim, J. L., Esteves Ferreira, M. J.: The Determination of Safety Factor of Arch Dams by Means of Models. L.N.E.C., Memoria Nr. 163, Lisbon (1961).

[16] Rosanov, N.: Etudes sur modèles élastiques de la statique des ouvrages hydrau- liques. Symposium on Concrete Dams Models, L.N.E.C., Paper Nr. 10, Lisbon (1963).
[17] Rowe, R. E.: Tests on Four Types of Hyperbolic Shell. Proceed. of the Symposium on Shell Research, Delft (September 1961).
[18] Serafim, L., Cruz Azevedo, M.: Methods in Use at the L.N.E.C. for the Stress Analysis in Models of Dams. L.N.E.C., Memoria Nr. 201, Lisbon (1963).
[19] Serafim, L., Poole da Costa, J.: Methods and Materials for the Study of the Weight Stresses in Dams by Means of Models. Rilem, Intern. Colloq. of Mod. of Struc- tures, Madrid (1959).
[20] Templin, R. L.: Tests of Engineerings Structures and their Models. Transactions, ASCE, Vol. 102, 1211—1225 (1937).

4.   Linearly-Elastic Static Models

4.1   General Remarks

It has been shown that experiments on models of category 2 (see section 1.4) reflect conceptually the traditional mathematical procedures for an elastic analysis of the whole structure.

As more often the theoretical analysis gives incomplete results or requires an approximate idealisation, so experimentation on a model becomes attractive in that it constitutes a more complete and reliable investigational tool capable of taking into account the structural stiffnesses in the linear elastic range in their entirety and with their true values.

The elastic model provides in particular the values of the force resultants transmitted by individual structural elements (beams, columns, portals, screen walls, connecting joints, etc.), these being necessary for verifying the safety of locally more highly loaded sections and also for eventually dimensioning the reinforcement.

In particular cases the elastic experiments on a model of the entire structure can be followed by investigations to failure on microconcrete models of the individual structural elements listed above: tests to failure that are performed by applying and proportionally increasing up to collapse the forces obtained from the elastic model.

Logically the elastic model investigations can proceed in parallel with and in conjunction with theoretical analyses.

The model can frequently provide information in areas too difficult or complex to be treated analytically (settlement of foundation, influence lines, etc.), which can be used to develop a more complete and reliable theoretical analysis. More generally the two methods (experimental and theoretical analysis) normally provide a set of results varying in extent and accuracy so as to constitute methods of investigation that frequently complement or confirm each other by turn.

For the reasons given above and because of their simplicity of conception and application, elastic models are those most widely used in universities, laboratories and by design offices. A minimum of equipment and skill with the help of trained technical assistance makes the construction and use of these models relatively easy.

Nevertheless it should be pointed out that this type of model is of little use as a tool for investigating the real overall factor of safety of the project, in as much as only an experiment taken beyond the elastic limit and up to failure can establish the overall behaviour by calling into play in its entirety and up to collapse the load bearing capacity of the entire structure.

On the other hand elastic models lend themselves to dynamic experiments for determining oscillatory effects induced by variable wind loading (suspension bridges) and the consequences of seismic effects (bridges, skyscrapers, etc.); these latter being topics beyond the scope of the present publication.

4.2 Principles of Reciprocity and Superposition (Model of the Parque Central Skyscraper, Caracas)

Because the models at present being discussed operate in the linear elastic range we can make use of the principles of reciprocity and superposition for the measurement of deformations and stresses.

These principles can be used, for example, for evaluating the effect of wind loading on the structure. The model of the Parque Central skyscraper in Caracas, of which several examples were reproduced, provides a suitable example in this context (Fig. 29).

Fig. 29. General view of the model of the Parque Central skyscraper, Caracas

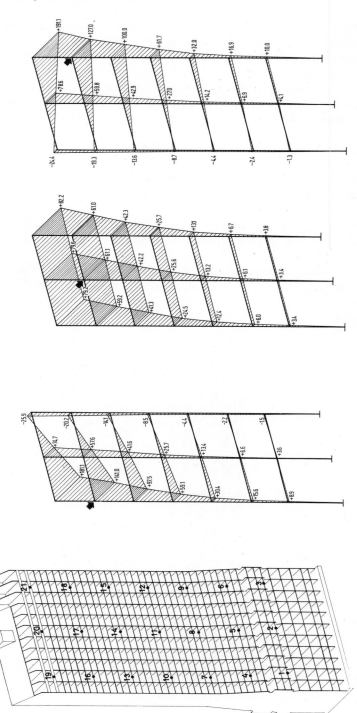

Fig. 30. Model of the Parque Central skyscraper, Caracas: influence surfaces for concentrated unit loads normal to the walls at points 16, 17, and 18

The structure basically consists of vertical wind resisting walls intersecting the entire transverse section of the building and connected by horizontal floor slabs. The walls are supported at the lower levels by box structures. The model was designed to reproduce schematically this basic structure at a scale $\lambda = 40$. The material used was epoxy resin filled with powdered silica and having an elastic modulus $E = 60,000$ kg/cm^2.

For the deformations, influence surfaces[4] were measured on the model relative to unit loads (2,000 ton in the prototype) applied to 21 points of the longitudinal walls as shown in Fig. 30.

As an illustration the influence surfaces for unit loads normal to the walls for points 16, 17 and 18 are shown in the same figure. The influence surfaces corresponding to 7 points along the middle axis of one of the end transverse walls were also determined.

At this point it was necessary to evaluate the normal pattern of loading to be applied to the model for different angles of incidence of the wind. The loading pattern to be applied was finally distributed to the n elementary volumes associated with the n points for which the surfaces of influence had been determined.

If N is the component of bending normal to point k for unit normal load applied at an arbitrary point h and P_h is the thrust relative to the elementary volume associated with the point h, deduced from the distribution of the loading pattern, then the total bending deflection at k for the given load pattern is given by the summation:

$$\sum_{1}^{n} N_{kh} P_h$$

The stress distribution can be dealt with in an analogous manner. On this point Fig. 31 shows the stresses on two transverse sections for unit loads (1,000 ton in the prototype) applied normally to point 20.

If we wish to check the structure for a significant number of hypotheses the use of a small computer facilitates the extensive summation processes.

In principle one can foresee the possibility of achieving, in the general case, a totally automatic experimental procedure in association with digital data loggers and computers. It must be realised however that automatic recording and analysis of results cannot go totally unmonitored since the possibility of instrumental errors necessitates a continuous check being kept by the operator on for example the influence surfaces, at least with regard to their consistency and plausibility.

It is often difficult to isolate the causes of errors from a critical examination of the completely processed results, so that it is frequently necessary to repeat all or part of the experimental programme.

---

4 In effect, in the present case the measurements of the influence surfaces amounted fundamentally to determining the "flexibility matrix" of the structure for the purpose of investigating the behaviour of the structure when subject to seismic disturbances. Nevertheless this matrix is ideally suitable for investigating the effect of wind loading on this structure.

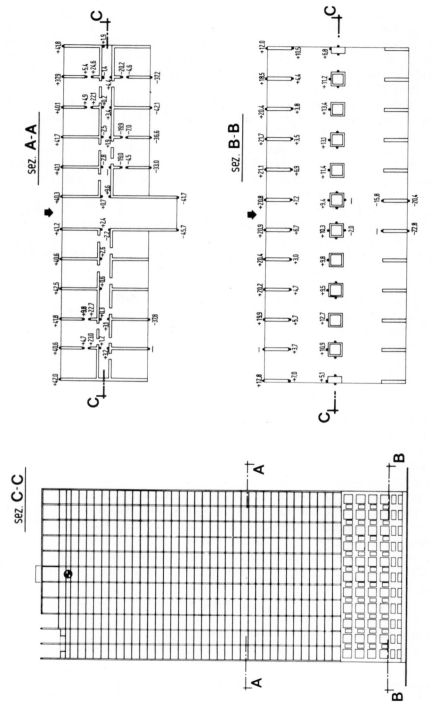

Fig. 31. Model of the Parque Central skyscraper, Caracas: stresses (kg/cm²) for a unit force normal to the wall at point 20

4.3 Model of the Roof of the Temple of the Blessed Virgin Mary

For complex structures it is generally so difficult to predict the load distribution due to wind action that it is frequently necessary to perform tests on a model in a wind tunnel.

Such was the case for the model of the Temple of the Blessed Virgin Mary (Trieste) (Fig. 32). Experiments were carried out in a wind tunnel on a model to a scale $\lambda = 100$ by the Institute of Aeronautics at the Polytechnic of Turin.

Fig. 32. Model of the Temple of the Blessed Virgin Mary, Trieste

Their reports gave the values of the following non-dimensional quantity (Fig. 33):

$$K = \frac{p - p_s}{\frac{1}{2} \frac{\gamma}{g} V^2}$$

that is of the increase of surface pressure $p - p_s$ with respect to the static pressure p_s referred to the value $\frac{1}{2} \frac{\gamma}{g} V^2$ ($\frac{\gamma}{g}$ = density of air, V = wind velocity), relative to a certain number of points on the external surface.

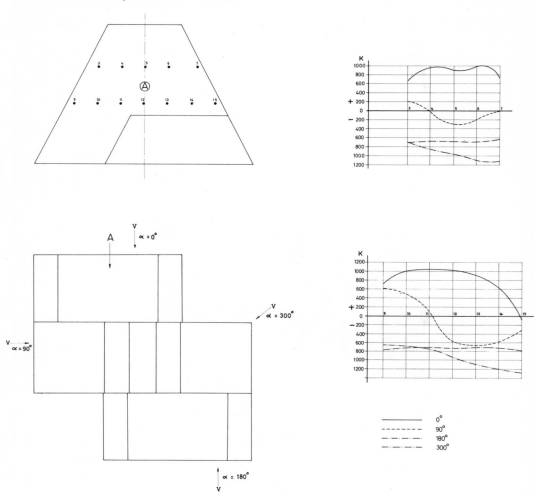

Fig. 33. Model of the Temple of the Blessed Virgin Mary, Trieste. Variation of K for different wind incidence angles

Note that K is taken as a constant, characteristic of a given point for a given incident wind direction, and does not depend on the intensity of the wind.

Knowing the local value of K the surface loads acting on the real structure at any individual point can be expressed in the form:

$$\Delta p = p - p_s = K\,{}^1\!\frac{1}{2}\frac{\gamma}{g}\,V^2$$

Referring to mean ambient meteorological data the following values were assumed, $\frac{\gamma}{g} = 0.125$ kg sec^2/m^4 (i.e. specific weight of air $\gamma = 1.23$ kg/m^3, $g = 9.81$ m/sec^2) and the wind velocity taken as 200 km/h, equal to approximately 55 m/s (Fig. 33).

The value of the surface wind load can therefore be expressed at a given point as a function of K in the form:

$$\Delta p = 190 \ K \ (\text{kg/m}^2)$$

From an examination of the results of the wind tunnel tests it appeared that the most critical wind loading situations were those corresponding to winds in the directions perpendicular and parallel to the principal vertical façades. The experiments were therefore performed for loading conditions relative to these two configurations.

The static model, constructed in celluloid, was tested at the scales $\lambda = 48$, $\zeta = 12$.

Considering the simplicity of the idealisation in this case recourse was not made to determining the influence surfaces, the required wind load distribution being applied directly to the model (Fig. 32).

4.4 Bridge Structures on the Rome-Florence Express Railway

Another model for which ample use was made of the principles of reciprocity and superposition was that connected with the Zorzi design for bridge and viaduct structures along the express railway between Rome and Florence.

The model was constructed at scales $\lambda = 20$, $\zeta = 5$ in filled epoxy resin (Fig. 34).

Since the model had two planes of symmetry, measurements were concentrated on a quarter of the structure. Stresses and deformations were readily determined for a whole range of load conditions using the principle of superposition. The symmetrical behaviour of the structure was checked by making appropriate measurements on the three remaining quarters of the model.

Fig. 34. Model of a deck beam for the Rome—Florence express railway

The principal objective of the investigation was to determine the interaction effects between the two prestressed beams for an eccentric load (train on one track only) with measurements of the twisting action on the beams and of the resulting inclination of the plane of support of the rail tracks.

4.5   Experiments in the Elastic Range on a Model of a Skyscraper

The model, constructed in celluloid, reproduced at scales $\lambda = 52.8$, $\zeta = 13$ the Borse skyscraper in Victoria Place, Montreal (Fig. 35) [21].

For the reproduction of the load bearing elements of the structure in the model the following idealisation criteria were adopted: the dimensions of the columns and of the central cross vault were varied, with increasing height, in a discontinuous manner (with discontinuities related to three floors associated with the technical services); the floor structures were realised with slabs of constant thickness so as to correctly reproduce the flexural rigidity of the floors of the project. The reinforcement called for in the floor in the region of the columns was produced by locally increasing the thickness of the floor slab.

Fig. 35. Model of the Victoria Place skyscraper, Montreal

The model was founded, through a celluloid plinth, on a 30 cm thick foundation block. This latter was made of araldite filled with pumice stone granules to reproduce the flexibility of the ground.

The behaviour of the structure for the effect of wind loading was investigated for two different loading schemes relative to two different values of the angle $\emptyset$ between the wind direction and the normal to a façade of the building.

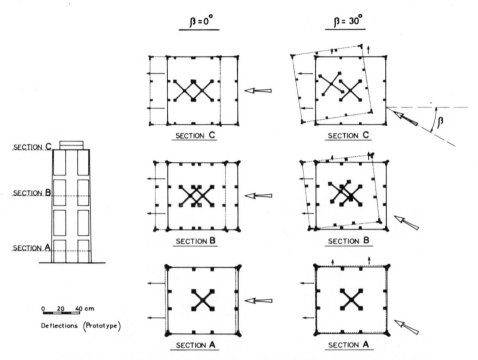

Fig. 36. Model of the Victoria Place skyscraper, Montreal: deflections at various levels of the building for different wind incidence angles

Again in this case the pressure distributions due to wind loading were found from wind tunnel tests.

The load pattern was distributed in 5 sections at different levels on two adjacent faces by means of steel load rods fixed to the floors and loaded by hydraulic jacks pulling against an external restraining structure.

The deflections obtained for $\emptyset = 0°$ and $\emptyset = 30°$ are shown diagrammatically in Fig. 36 for three sections of the structure.

10 and 20 mm gauge length electrical resistance strain gauges were used to measure stresses. For the central cross vault the electrical resistance strain gauges were applied as rosettes to enable the principal stresses to be determined at each measuring point.

Flat beam type deflection gauges were used to measure the deflections. The beams, deflecting to follow the deformations, transfer equal and opposite deformations to a pair of electrical resistance strain gauges on opposite faces of the beams. These strain gauge deformations are linearly proportional to the deflections of the beams.

Finally it is interesting to note that these elementary experiments showed that the structure, as originally designed, had too low a torsional moment of resistance. This evidence suggested some modifications with the object of increasing the torsional stiffness at the three levels associated with the technical services.

4.6   Elastic Model of the Pier of a Viaduct

This model, at a scale $\lambda = 50$, was of a pier of the Polcevera, Genoa, viaduct designed by Morandi and was constructed of celluloid ($\zeta = 12$) (Fig. 37) [21].

The structure consisted of a long deck beam supported in the centre by a pier and at each end by two large inclined tension bars fixed to the upper end of the central pier.

Fig. 37. Model of a pier of the Polcevera viaduct, Genoa

The model was tested in the elastic range for various conditions of load (dead loads and live loads) and for the effect of wind loading.

The loading system consisted of calibrated weights or hydraulic jacks acting on the deck beam through wooden pads with cork soles.

Deflections (through dial gauges) and stresses (by means of mechanical extensometers and electro-acoustic strain gauges) were measured for the following load conditions:

concentrated vertical loads acting on the deck beam (Fig. 38);

distributed load on a quarter, on a half and on the whole of the deck beam;

horizontal load (wind load) relative to a prototype load of 460 kg/m², applied to the deck, to the tension bars and to the central pier.

Finally dynamic tests were carried out to determine the more important resonant frequencies of the structure.

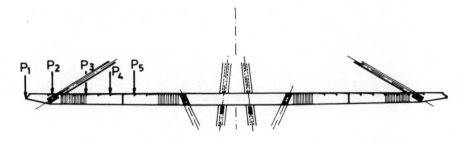

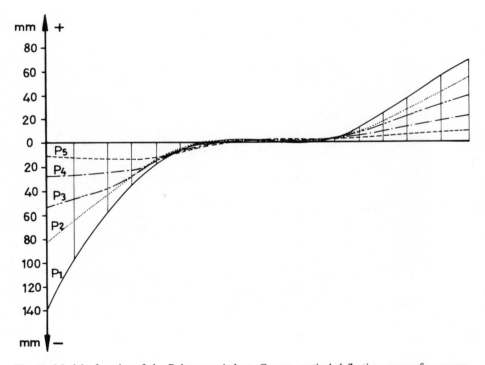

Fig. 38. Model of a pier of the Polcevera viaduct, Genoa: vertical deflection curves for concentrated loads applied at different point on the bridge deck

4.7   Elastic Model of a Spherical Domed Roof

This elastic model at a scale $\lambda = 50$ was of the roof of the Nervi design for the Cultural Centre of Norfolk (Virginia, U.S.A.). It was constructed from a material based on araldite and sand, reproducing the mechanical properties of the prototype (concrete) at a scale $\zeta = 5$ (Fig. 39).

Fig. 39. Model of the roof of the Cultural Center, Norfolk, Virginia

Fig. 40. Model of the roof of the Cultural Center, Norfolk, Virginia: disposition of the small distribution beams for the vertical loads

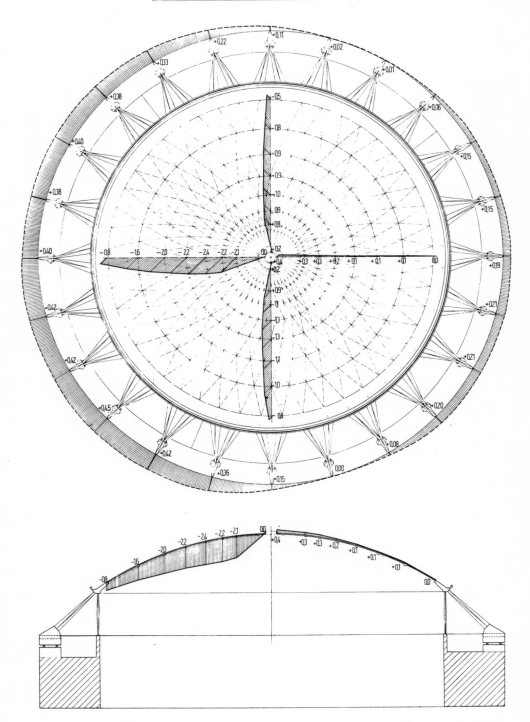

Fig. 41. Model of the roof of the Cultural Center, Norfolk, Virginia: displacement curves for
a uniform vertical load on half the dome

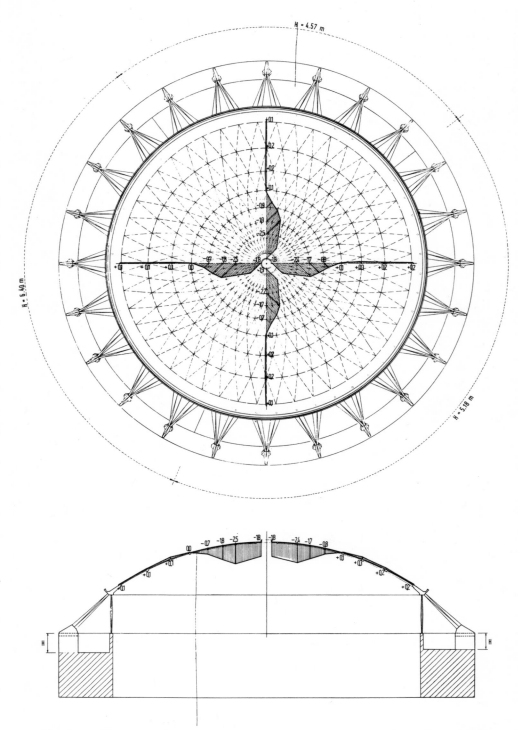

Fig. 42. Model of the roof of the Cultural Center, Norfolk, Virginia: displacement curves for
a uniform vertical load applied to a restricted central area of the dome

The real structure consists of a spherical dome with internal ribbing in rein-forced concrete, contained by a ring beam and supported by inclined columns that follow the curfature of the dome on to a strong foundation ring of prestressed concrete.

The experiments on the model were designed to study the static behaviour of the structure under the action of self weight loads and surface loads (symmetric and assymetric) using the loading equipment already detailed as an example in Fig. 16 and for which Fig. 40 shows the disposition of the small balanced load distribution beams.

Fig. 41 and 42 shows diagrammatically the deflection patterns for loads limited respectively to half the dome and to a rather restricted central area.

4.8 Gerber Designed Beam for an Autostrada Viaduct

The model, shown in Fig. 43, was investigated for the purpose of determining the feasibility of a deck beam suitable for general use for multiple span structures with individual spans of 40 m. The celluloid model was reproduced at a scale of $\lambda = 26.6$.

The deck beam was designed around six longitudinal beams (three for each carriageway), with small transverse connecting beams.

Fig. 43. Model of a Gerber beam for an autostrada viaduct

The investigations explored in particular, through measurements of deflections and stresses, the mutual interaction between the successive spans as well as the transverse behaviour of the structure for eccentric loads and the resulting torsional action induced in the longitudinal beams.

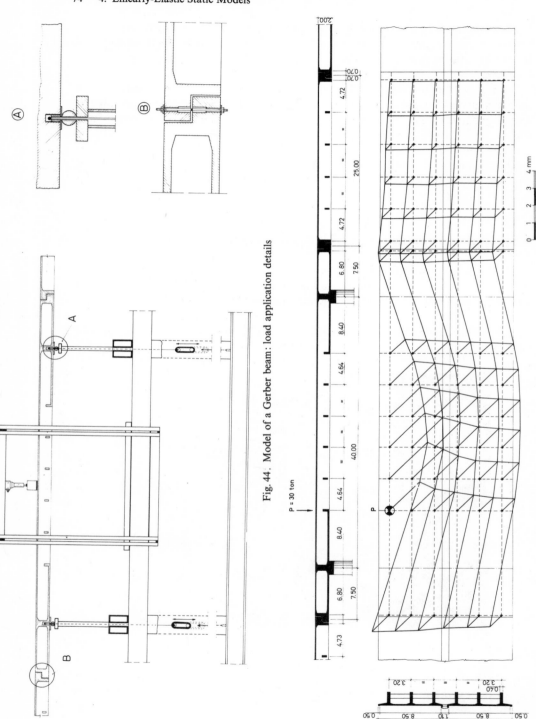

Fig. 44. Model of a Gerber beam: load application details

Fig. 45. Model of a Gerber beam: deflection pattern resulting from a concentrated load (30 ton in the prototype) applied at one point of the structure

To prevent the beam lifting from its supports in the absence of self weight with consequent distortion of the results, the supports were constructed from hollow cylinders through which passed steel wires loaded by ring dynamometers (Fig. 44). This system applied the necessary forces to maintain contact without increasing the stiffness by encastration. Also the support points of the intermediate joints were specially designed to produce a non restraining contact.

Measurements were made by determining the influence lines for a unit load at various points on the structure. For this purpose a frame capable of being moved along the model provided the reaction for a hydraulic jack which in turn could be moved transversely across the model (Fig. 44).

Fig. 45 shows, as an illustration, the deformed surface for a unit load (30 ton in the prototype) applied to one point of the structure.

4.9   Experiments by the L.N.E.C. of Lisbon

The following sub-sections give examples of experiments on elastic models by the "Laboratorio Nacional de Engeharia Civil" of Lisbon.

4.9.1   Tagus Bridge

This concerns the famous suspension bridge on the Tagus. This has an overall length of 2300 m and a central span of 1016 m.

Two models were constructed at a scale $\lambda = 50$ reproducing two stages of utilisation of the bridge, as a road bridge and as a mixed road-railway bridge (Fig. 46). The piers and the deck beam were reproduced in perspex and the cables by steel wires.

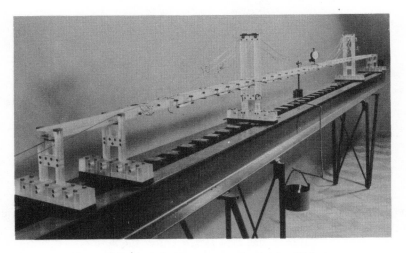

Fig. 46. Model of the Tagus suspension bridge, Lisbon

The internal stress resultants for different load cases were measured by means of resistance strain gauges. The results of the model studies were compared with results obtained for matrix calculations of the force distribution in the cables; these calculations were generalised to take account of the non-linear behaviour of the cables.

4.9.2 Tower Building

The model reproduced, at a scale $\lambda = 30$, a 23-storey reinforced concrete building of square plan form (Fig. 47).

The floor slabs were reproduced with steel plates and the walls from perspex sheets variably perforated so as to reproduce the varying thicknesses of the walls.

Fig. 47. Model of a tower building

The deformability characteristics of the foundation were also reproduced in the model.

As shown in the figure it was also tested under the action of dynamic loads applied to the foundation.

4.9.3 Tomar Bridge

This is a three span bridge of respectively 14, 48, 13 m in prestressed concrete.

The model was constructed from perspex at a scale $\lambda = 50$ (Fig. 48) from which were determined the influence lines for loads moving in longitudinal and transverse directions.

Fig. 48. Model of the Tomar bridge

4.10 Model of Milan Cathedral

Investigations have been carried out by means of an elastic model in relation to the non-uniform settlements occurring in the foundations of Milan Cathedral movements resulting from dewatering of underlying water bearing strata) and to the consequent damage to the delicate equilibrium of the structure. The model reproduced the entire structure of the "Tiburium" which is carried on 16 cross vaulted columns and supported on the same. The model was reproduced at scales $\lambda = 15$, $\zeta = 4.8$ (Fig. 49) [22].

To take account of the different deformabilities of the materials used in the mediaeval construction, those used in the model were separated into three basic categories with the following properties:

	Elastic modulus of the prototype
Brick walls	100,000 kg/cm^2
Granite (Serizzo) of schistose texture	240,000 kg/cm^2
Candoglia marble	570,000 kg/cm^2

The materials used in the models were epoxy resins filled in varying proportions with sand and polystyrene grains.

Fig. 49. Model of the Tiburium of Milan Cathedral

The actual structure of the "Tiburium" is essentially of bricks (Fig. 50) with the incorporation of four granite arches. The columns were built with an external facing of Candoglia marble on a granite core.

In the model the columns were placed on adjustable supports so as to permit translation and rotation and consequently the study of the influence of such movements on the equilibrium of the whole structure.

The self weight loads were applied to the model by means of rubber ring dynamometers and a movable anchor plate operated by a battery of 12 hydraulic jacks.

A multiplicity of tests were carried out for different displacements at the bases of the columns with measurements of the stress distributions in the arches and in the same columns. The principal purpose of these tests was to interpret the mode of static behaviour occurring in the structure and to isolate the cause of cracks that have been discovered.

The influence of the restraints for the thrust of the sharply pointed arches was examined on the models before proceeding to dimensioning and positioning them in the work.

Fig. 51 shows a plan view of the supporting columns of the "Tiburium". In the model the column shown shaded was given a unit settlement (equivalent to 1 cm settlement in the real structure).

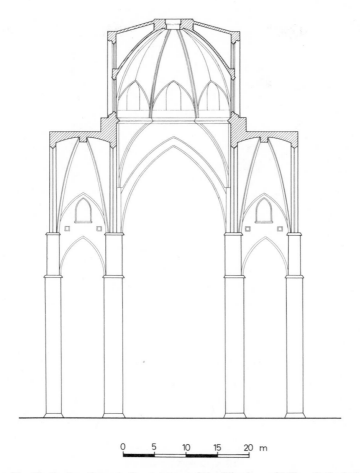

Fig. 50. Section through the structure of the Tiburium of Milan Cathedral

Putting the resulting decrease in load on this column equal to 100 the diagram shows the percentage variation in load (increase or decrease) found in the other columns.

From this set of results, using the principles of reciprocity and superposition, it is possible to impose different settlements on the 4 central columns and then evaluate the various distributions of loads on the central and peripheral columns.

In the same way by imposing a unit settlement separately to two of the peripheral pillars (one corner and one intermediate one) their effect on the percentage load variation in the other pillars was evaluated. Thus by symmetry and using the principles referred to above, the redistributions of loads for any pattern of settlements of any or all the pillars under examination can be evaluated with the aid of a computer.

For the behaviour to failure of single columns the reader is referred to section 5.2.5.

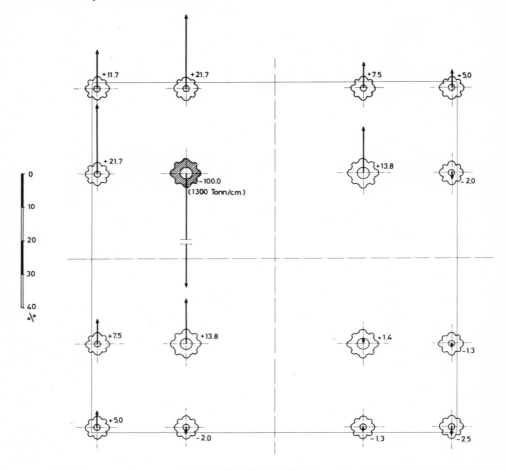

Fig. 51. Model of the Tiburium of Milan Cathedral: diagram showing the percentage variation of load in the columns following unit settlement (1 cm in the prototype) of one of the four central columns

4.11 Determination of the Effect of Wind Loading on a Hyperbolic Paraboloid Roof

This model, constructed by the "Centre Experimental de Recherches e d'Etudes du Batiment et des Travaux Publics" of Paris, reproduced at a scale $\lambda = 125$ the hyperbolic paraboloid shell roof of the French Pavilion at the 1958 World Exhibition in Brussels.

Support for the roof was achieved in the prototype by stretched steel cables placed as two orthogonal families of curves and anchored to robust reticular beams located along the perimeter of the structure.

Fig. 52. Shell roof of the French Pavilion at the Brussels Exhibition: model tested in the wind tunnel without the side walls

The particular feature of the brass model produced was that it was tested directly in a suitable tunnel, at a wind speed of 25 m/sec, firstly without the side walls (Fig. 52) and subsequently after these walls had been added (Fig. 53).

The stresses in the cables and in other elements of the structure were measured with electrical resistance strain gauges.

Fig. 53. Shell roof of the French Pavilion at the Brussels Exhibition: model tested in the wind tunnel with side walls added

References

[1] Balint, P. S., Shaw, F. S.: Structural Model of the "Australia Square" Tower in Sydney. Architectural Science Review, Sydney, Vol. 8, Nr. 4, 136—149 (December 1965).

[2] Beggs, G. E.: The Use of Models in the Solution of Indeterminate Structure. Franklin Institute, Vol. 203 (1927).

[3] Beggs, G. E.: Test of a Celluloid Model of the Stevenson Creek Dam. Arch Dam Investigation, ASCE (1928).

[4] Billington, D. P., Mark, R.: Small Scale Model Analysis of Thin Shells. ACI Journal, Proceedings, Vol. 62, Nr. 6, 673—688 (June 1965).

[5] Breen, J. E.: Fabrication and Tests of Structural Models. Proceedings ASCE, Vol. 94, Nr. ST6, 1339—1352 (June 1968).

[6] Carpenter, J. E.: Structural Model Testing. Compensation for Time Effect in Plastics. Journal, PCA Research and Development Laboratories, Vol. 5, Nr. 7, 47—61 (January 1963).

[7] Carpenter, J. E., Magura, D. D., Hanson, N. W.: Structural Model Testing-techniques for Models of Plastic. Journal, PCA Research and Development Laboratories, Vol. 6, Nr. 2, 26—47 (May 1964).

[8] Charlton, T. M.: Direct Method for Model Analysis of Structures. Civil Engineering and Public Works Review, London, Vol. 48, Nr. 559, 51—53 (January 1953).

[9] Coull, A.: Tests on a Model Shear-wall Structure. Civil Engineering and Public Works Review, London, Vol. 61, Nr. 722, 1129—1133 (September 1966).

[10] Coull, A., Das, P. C.: Analysis of Curved Bridge Decks. Proceedings, Institution of Civil Engineers, London, Vol. 37, 75—85 (May 1967).

[11] Cowan, J. H.: Some Applications of the Use of Direct Model Analysis in the Design of Architectural Structures. Australian Building Research Congress, Sydney (1961).

[12] Fialho, J. F. L.: The Use of Model Tests in Bridge Analysis. Seventh Congress of the International Association for Bridge Engineering, Rio de Janeiro (1964).

[13] Fumagalli, E.: The Use of Models of Reinforced Concrete Structures. Magazine of Concrete Research, Vol. 12, Nr. 35, 63—72 (July 1960).

[14] Guralnick, S. A., La Fraugh, R. W.: Laboratory Test of a 45-foot Square Flat Plate Structures. ACI Journal, Proceedings, Vol. 60, Nr. 9, 1107—1186 (September 1963).

[15] Haddon, J. D.: The Use of Wind-Tunnel for Determining the Wind Pressure on Buildings. Civil Engineering and Public Works Review, London, Vol. 55, Nr. 645, 500—502 (April 1960).

[16] Hatcher, D. S., Sozen, M. A., Siess, C. P.: A Study of Tests on a Flat Plate and a Flat Slab. Civil Engineering Studies, Structural Research, Series Nr. 217, University of Illinois, Urbana (July 1961).

[17] Lauddeck, N. E.: A Direct Method for Model Analysis. Proceedings ASCE, Vol. 82, Nr. ST1 (January 1956).

[18] Lim, B. P.: Experimental Stress Analysis of a Shallow Spherical Dome, with Particular Reference to its Modes of Support. Constructional Review, Sydney, Vol. 38, Nr. 2, 18—29 (February 1965).

[19] Little, G., Rowe, R. E.: Load Distribution in Multi-webbed Bridge Structures from Tests on Plastic Models. Technical Report TRA, Nr. 185, Cement and Concrete Association, London (May 1955).

[20] Nervi, P. L.: Structures. New York: F. W. Dodge Corp. 1956.

[21] Oberti, G.: La ricerca sperimentale su modelli strutturali e la Ismes. „L'Industria Italiana del Cemento", Anno XXXIII, Nr. 5 (May 1963); Bulletin Ismes Nr. 22 (January 1964).

[22] Oberti, G. Model Analysis for Structural Safety and Optimization. Bulletin Ismes, Nr. 41 (February 1970).

[23] Rocha, M.: Structural Models. The State of the Art. Conference, Preprint Nr. 473, ASCE (May 1967).

[24] Rocha, M.: Model Tests in Portugal — Part 1 and 2. Civil Engineering and Public Works Review, London, Vol. 53, Nr. 619, 49—53 (January 1958).

[25] Rowe, R. E.: Works on Models in the Cement and Concrete Association. Journal PCA Research and Development Laboratories, Vol. 2, Nr. 1, 4—10 (January 1960).

[26] Rowe, R. E., Base, G. D.: Model Analysis and Testing as a Design Tool. Proceedings, Institution of Civil Engineers, London, Vol. 33, 183—199 (1966).

[27] Stevens, L. K.: Investigation on a Model Dome with Arched Cutouts. Magazine of Concrete Research, London, Vol. 11, 3—14 (1959).

[28] Thompson, J. C., Godden, W. G.: Experimental Study of Model Tied-arch Bridge. Institution of Civil Engineers, London, Vol. 14, 383—394 (December 1959).

[29] Villard, A.: Investigations on Reduced Models of Reinforced Concrete Structures. Bulletin Tech. Suisse Rom., Vol. 79, Lausanne (June 1953).

[30] Westergaard, H. M., Slater, W. A.: Moments and Stresses in Slabs. Proceedings, ACI, Vol. 17, 415—538 (1921).

[31] Wilson, W. M.: Laboratory Tests of Multiple-span Reinforced Concrete Arch Bridges. Asce, Vol. 100, 424—454 (1935).

[32] Model Testing. Proceedings of a One-day Meeting held in London on 17 March 1964, Cement and Concrete Association, London (1964).

5. Model Simulation of Reinforced and Prestressed Concrete Structures

5.1 Introduction

The model simulation of reinforced concrete at reduced scale for $\zeta = 1$ does not normally pose particularly complex problems.

It has been shown that a cementaceous microconcrete, with suitably reduced aggregate dimensions and manufactured with care and skill can satisfactorily reproduce an ordinary concrete, always taking account of the range of properties that can occur in concrete in the normal course of events[5].

It is necessary however to be aware of the limits of applicability imposed by this type of model simulation for massive structures in which self weight loading is significant.

For structures for which the application of self weight loading is onerous and difficult it is sometimes convenient to resort to a model simulation with $\zeta > 1$ in accordance with equation (10).

It is important to note here that it is difficult to find a material that satisfactorily reproduces the particular properties of steel for any ratio $\zeta > 1$[6].

We can however resort to the device of reducing the area of the cross section of the steel in the ratio

$$\frac{\Omega}{\Omega'} = \zeta \lambda^2$$

The device is equivalent in practice to using in a normal structure a hypothetical steel that, for the same extension at failure, develops a specific resisting

---

5 We have seen how the presence of obstructions determines the maximum permissible dimensions of the aggregate. It is however advisible to make sure that the minimum thickness of the model (slabs and ribs) is not less than 8 to 10 mm. Such a minimum thickness is determined both by practical construction requirements and by the need to use aggregates with D max $\geqq 2$ mm. This requirement of necessity limits the geometrical scale reduction λ. For example for the models of the Pirelli skyscraper (5.3.4) and the Cathedral of San Francisco (5.2.3) (Fig. 61) this factor led to a reproduction scale of $\lambda = 15$.

6 For completeness the use of aluminium or copper wires and in one case nylon filaments should be noted. Their use is perfectly acceptable for experiments within the elastic range, much less so for experiments taken to failure.

force ζ times greater and requires its cross section to be reduced by the same factor.

The total strength contributed by the steel is not changed, but there is a reduction in the bonding surface of the steel by a factor $\sqrt{\zeta}$.

The area of this surface is in practice further reduced when for practical convenience in the model the number of tendons is reduced, with a corresponding increase in diameter to maintain the required total section.

We note that such idealisations must sometimes of necessity be used even in models where $\zeta = 1$.

A further precaution consists of ascertaining that small diameter sections display the same extension and strength properties as normal reinforcing steel. Normally small diameter sections produced by wire-drawing have inferior properties with an appreciably reduced extension at failure. It is advisable in this case to arrange for a check to be made on the properties of the steel to ensure that they fall within suitable limits.

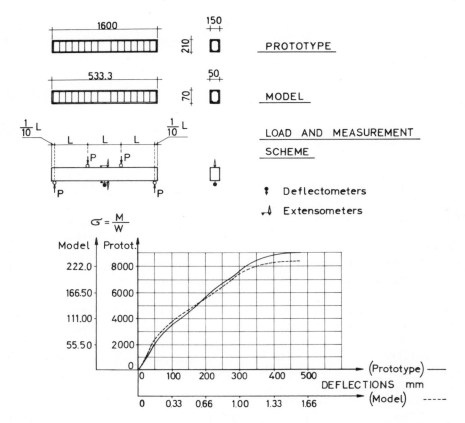

Fig. 54. Comparison tests on reinforced concrete and reinforced pumice beams for $\lambda = 3$, $\zeta = 4$

Joints between reinforcement members can be made satisfactorily with tin solder.

To determine the influence of reduced bond strength we have carried out flexural tests on small reinforced beams in concrete and pumice mortar respectively, these latter to scale ratios:

$$\lambda = 3, \; \zeta \simeq 4$$

(see Fig. 54) [11].

For corresponding deformations the number of fissures in the small pumice beams was reduced in the ratio of about $^2/_3$ (Fig. 55).

Fig. 55. Comparison of crack patterns after bending tests

The deflection curves of model and prototype (Fig. 54) are in good agreement up to 80% of the failure load, that is to say up to the onset of fissuration; after which the curve for the model beam falls off more rapidly showing finally a reduction in strength of approximately 7%. For higher ratios of efficiency ζ this discrepancy is accentuated.

It is worth noting however that the reduction in strength resulting from a reduced bond strength in the model works, in every case, in favour of greater safety for the real structure.

Also no problems exist in the reproduction of prestressing cables, again in this case making exception for failure to reproduce bond conditions. It is useful in this context to note that we do not normally grout the cables in models so that if necessary their tension can be changed or, when required, we can measure their variation in tensions, as for example in the case where large deformations occur and at the onset of failure.

When modelling structures in bricks or stone masonry it is advisable, whenever possible, to use these same materials, that is to say to resort to a model for which $\zeta = 1$.

5.2 Model Studies with Ratio of Efficiency $\zeta = 1$: Some Classical Examples

5.2.1 Floor Slab of the Pirelli Skyscraper

Section 5.3.4 gives details of a model of the entire structure and the reader is referred to this for more complete information.

The structure of a prestressed concrete floor slab was alternatively reproduced separately in a model to scale $\lambda = 5$ (Fig. 56) [23]. The material used was a fine concrete with aggregate for which $D_{max} = 6$ mm possessing the following properties:

$$E = 300,000 \text{ kg/cm}^2 \qquad \sigma_{uc} = 350 \text{ kg/cm}^2$$

Fig. 56. Model of a floor of the Pirelli skyscraper (Milan)

The section of the floor slab reproduced consisted of 3 ribs prestressed with longitudinal cables and the associated decking. It was limited to two of the three longitudinal spans of the design, i.e. the central span and a lateral one (Fig. 57).

The external end of the small lateral span was encastered into a rigid vertical wall which itself was constrained above and below to reproduce the stiffening effect contributed by the vertical structure enclosing the ends (Fig. 86). The free end of the central span was simply supported but was extended in a springboard

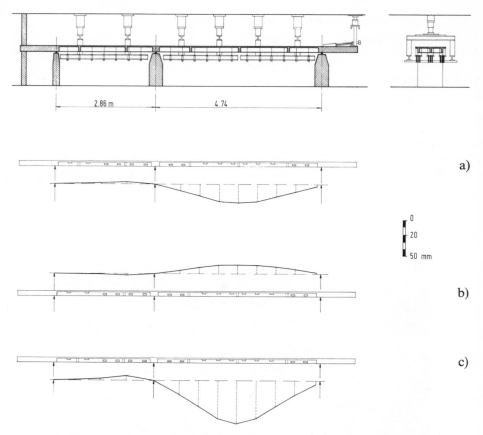

Fig. 57. Floor of the Pirelli skycraper. Drawing of the model showing deflection patterns: a) uniformly distributed load of 400 kg/m² without prestressing, b) each joist prestressed to 30 t, c) uniformly distributed load of 800 kg/m² with the addition of prestress

like appendage to which corrective loads were applied by hydraulic jacks. During the loading tests these jacks provided the forces necessary to maintain symmetrical deflections in the central span, required to reproduce the action of the missing smaller lateral span.

The live loads were applied by a standard system of ring dynamometers with attachment points distributed uniformly on the upper surface of the decking.

Loading was achieved through a battery of hydraulic jacks acting on a rigid metallic distribution plate.

In order to reproduce the stages of construction and the loading programme of the prototype, the model was tested under the following conditions:

a) floor slabs subject to self weight and to the temporary overload due to the action of supporting (propping up) the floor above during the process of casting and curing this latter, no prestressing loads present (intensity of this load 400 kg/m²);

b) pretensioning of the cables up to 30 ton for each rib beam;

c) application of the total load (dead and live) of magnitude 800 kg/m² on the prestressed floor slab.

Measurements of stresses and deformations were made in the zones of greatest interest during experiments in the normal working range.

Deflection curves relative to loading conditions a), b) and c) are shown in Fig. 57.

The model was finally loaded to failure.

The rib beams underwent considerable deflections and became highly fissured, finally carrying the load in catenary. The total load was of the order of 2,500 kg/m² at which the beams, without fully failing, were certainly in the large deformation range.

5.2.2 Model of a Prestressed Concrete Shell Roof (Centro Euratom-Ispra)

The model reproduced at a geometrical scale $\lambda = 5$ two elements of a folded triangular plate roof designed in prestressed concrete.

Fig. 58 shows the disposition of the reinforcement and the prestressing cables before the model was cast.

The surface loads were applied by the classical system of ring dynamometers.

Fig. 58. Model of a prestressed concrete roof (Euratom Building, Ispra). Positions of the steel reinforcement and the prestressing cables

The model, supported on a tubular framework, was tested at a load of 400 kg/m² after tensioning the prestressing cables to 10,000 kg/cm². The load was then increased to 800 kg/m² at a prestressing cable tension of 13,000 kg². Finally the test was taken to failure increasing the load to 1,600 kg/m², at which value an extensive system of cracks had developed in the model (Fig. 59).

Fig. 59. Model of a prestressed concrete roof showing fissuration after the test to failure

5.2.3 Model of San Francisco Cathedral

The model reproduced in elevation the architectural structure of the Cathedral consisting of 8 near vertical surfaces of hyperbolic paraboloid form, bonded on two sides to the framework forming the fundamental cross vault of the structure [25].

This complex was supported on near horizontal plates that transferred the load to four corner pillars. The entire structure was designed in normal reinforced concrete.

The model was reproduced to a scale $\lambda = 15$ in concrete with an aggregate having $D_{max} = 4$ mm. The material had the following properties:

$$E = 240,000 \text{ kg/cm}^2, \quad \sigma_{uc} = 380 \text{ kg/cm}^2.$$

Construction of the formwork and preparation of the metal reinforcement for the shells required particular care and accuracy because of their limited thickness (Fig. 60).

Fig. 60. Model of San Francisco Cathedral. Position of the steel reinforcement

Fig. 61. Model of San Francisco Cathedral during testing

The loads applied to simulate self weight and the weight of the superstructure were 230 kg/m² on the surfaces of the near-vertical plates and 300 kg/m² on the surfaces of the near-horizontal plates.

Application of the load was again acheived in this case by means of jacks, ring dynamometers, hydraulic jacks and movable anchor plates (Fig. 61).

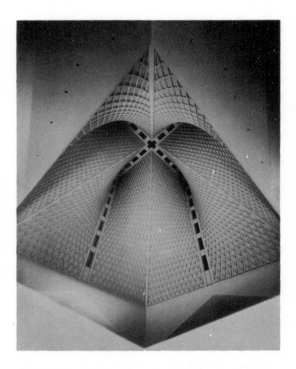

Fig. 62. Model of San Francisco Cathedral. General view of internal surface of the roof

The measurements of the stresses along the slender interlaced ribs on the internal faces of the near-vertical surfaces were carried out particularly accurately in the experiments that were performed in the elastic range (Fig. 62). This was necessary to determine the static behaviour of the structure and to compare it with the results of complex analytical calculations.

The tests in the elastic range were followed by tests to failure during which fissuring of the middle sections of the near-horizontal plates was examined, and more precise attention was given to the support of the ribs of the vertical cross.

As a result of these tests it was decided to replace the sub-horizontal plates with box structures containing internal diaphragms. Also increasing the section of the box structures in the region of the pillars substantially reduced the shearing action at the pillars.

A new analysis was then carried out on a second elastic model of the structure as modified, the model being constructed from epoxy resin to a scale $\lambda = 40$. The self weight tests were repeated on this model and additionally tests with wind loading were performed. The wind loads were evaluated from tests on an aerodynamic model tested at varying angles of incidence in a wind tunnel.

The test results were completely satisfactory.

5.2.4 Model of a Hyperboloid Shell in Bricks and Reinforced Concrete

The model investigated was of a roof which in plan view formed a square of approximately 10 m side. The roof was formed from four hyperboloid surfaces and was supported at the four corners (Fig. 63) [17].

The roof was of mixed construction of "terra-cotta" intersected with ribs of reinforced concrete and was reproduced at a scale $\lambda = 5$.

The brickyard produced purpose made "terra-cotta" models of the prototype components at the correct scale.

Fig. 63. Model of a hyperboloid shell. General view of the model during testing

The purpose of the tests was to analyse the behaviour of the whole structure constructed from heterogeneous materials, when subjected to the effect of dead and live loads which were assumed in the design to total 350 kg/cm^2.

The loading was applied by a system of loading rods, ring dynamometers, movable anchor plate and hydraulic jacks.

One of the more interesting aspects of the research was the dimensioning of the edge beams in relation to the forces transmitted by the shell. Using only one model it was possible to incorporate and analyse three different sizes of edge beams in the search for the optimum static solution.

The Achilles heel of the structure was shown to be at the supports where the stress concentration caused the local premature collapse of the structure (Fig. 64). As a result the final design was suitably locally reinforced both by increasing the quantity of steel and the effective concrete section.

Fig. 64. Model of a hyperboloid shell after the test to failure

5.2.5 Columns of Milan Cathedral

As an example of modelling a stone structure we can take, in completion of the model investigations of the Tiburium of Milan Cathedral referred to in section 4.10, the tests carried out on the columns of the same Cathedral which were reproduced at a scale $\lambda = 4.7$ [25].

For these models the same materials as those used in the prototype were employed: Candoglia marble for the external covering, granite (Serizzo) for the core.

The construction of the model followed the same techniques of craftsmanship employed by the Venerated Builder of the Cathedral, proceeding from the construction of the blocks to their assembly in columns with lime mortar (Fig. 65). The pillars were tested to failure under a load of 2,000 tons (Fig. 66).

The following conclusions were drawn from the results of the tests:

a) even though the two stone materials have approximately the same strength, of the order of 1,000 kg/cm^2, the stress distribution within the column is determined by their different moduli of elasticity. As already shown for the model of the Tiburium, the modulus of the marble is 2.4 times greater than that of the granite.

Fig. 65. Model of a column of Milan Cathedral during construction

Fig. 66. Model of a column of Milan Cathedral after the test to failure

The external covering of marble was therefore much more highly loaded so that it failed relatively early as shown by vertical fissuring and separation of the marble from the granite core.

This rupture process reproduced, at the stage of incipient fissuration, conditions observed in-situ.

b) the pillars followed the rule of thumb for modern technique, that is to say that perfectly finished, mounted and sealed surfaces have up to three times the strength of those produced by the more rudimentary methods of the middle ages. This accounts for the premature fissuration occurring in-situ.

All this has made practical the repair and restoration of the damaged areas with more modern and refined methods.

Confirmation of this was also obtained from model tests performed to check the repair methods. The model served as an excellent test bed for the chiselling work to be carried out in-situ.

5.2.6 Static Tests on Models of Prestressed Concrete Pressure Vessels for Brown Bovery Krupp (Mannheim)

Static investigations on models of prestressed concrete pressure vessels for nuclear reactors represent a new and advanced area of research of great interest in the development of methods of utilising nuclear energy.

The use of prestressed concrete for containing the core of a pressurised reactor basically satisfies two requirements:

one is that of utilising by means of prestressing a relatively cheap material such as concrete for containing reasonably high pressures.

the other is that of utilising the thickness of the concrete as an efficient biological shield against nuclear radiation.

The production of a high triaxial compression for a large pressure vessel poses new problems both with respect to traditional prestressed concrete technique, and to modelling methods.

Designing pressure vessels raises specific problems, in particular:

1. the study of the positioning and dimensioning of the prestressing cables, considering: the obstruction they cause, their frequently abnormal dimensions, the need for a uniform distribution of prestress in the concrete, etc.;

2. the analysis of pressure vessels suggests a biaxial and triaxial state of stress, thus highlighting the fact that the mechanical strength properties of concrete under these conditions are at present not fully understood, mainly because of the technological difficulties of testing.

Taken as a whole model tests represent a method of investigation, in the elastic range, capable of giving the stress and deformation distributions in parallel with and in confirmation of the results of mathematical analyses; but more significantly they provide a research tool for investigating the static behaviour of the structure in the visco-plastic range up to rupture, from which the overall factor of safety can be determined.

In the specific case the self weight can be assumed to be negligible making possible a model simulation with $\zeta = 1$.

As an example the tests carried out on a model pressure vessel for a THTR (Thorium High Temperature Gas Reactor) reactor will be described (Fig. 67) [12] [32].

Fig. 67. General view of the THTR reactor pressure vessel for Brown Bovery Krupp

The model, reproduced at a scale $\lambda = 20$, logically required an adequate schematisation and reduction in the numbers of the cables for applying the prestressing that were called for in the design (Fig. 68).

This schematisation, reproducing in the model the hypothetical use of high capacity cables, formed an interesting aspect of the research to define the admissible limit of concentration of the prestressing loads in a limited number of cables. This leads automatically to an investigation of the use of high capacity cables in the prototype.

Cables provided by BBR of Zurich were used for prestressing the prototype.

For the cylindrical body 7 mm dia. single strand cables were used, placed entirely within the casting according to a special ENEL patent disposition (Ing. Scotto) while the end plates used 30 multi-strand straight cables totalling 180 wires of 7 mm dia. The vertical prestressing was achieved by 36 cables each consisting of 4 wires of 7 mm dia.

Fig. 68. Model of the THTR reactor pressure vessel. Position of the prestressing cables

The dimensions of the model and the distribution of the reinforcement can be seen in Fig. 68.

The properties of concrete used were:

$$E = 400,000 \text{ kg/cm}^2, \ \sigma_{uc} = 600 \text{ kg/cm}^2.$$

Electro-acoustical and electrical resistance strain gauges were used to measure the strains. Overall displacements were measured with electrical inductance transducers (Hottinger) connected to an automatic digital data logger. Suitable calibrated dynamometers were used to control the pull in the prestressing cables.

Tests were performed on the model to determine the stresses and deformations due to the prestress in the cables as well as to the internal hydrostatic pressure.

To prevent hydraulic leaks during large deformation tests up to final failure (Fig. 69) the interior was sealed with a subsidiary copper sheet liner.

Fig. 70 shows on a radial section the final failure mode. This was basically caused by separation between the cylindrical body and the top end plate following failure of the vertical cables. This failure and the general state of fissuration shown occurred at an internal pressure of 191 kg/cm² compared with the normal design operating pressure of 40 kg/cm².

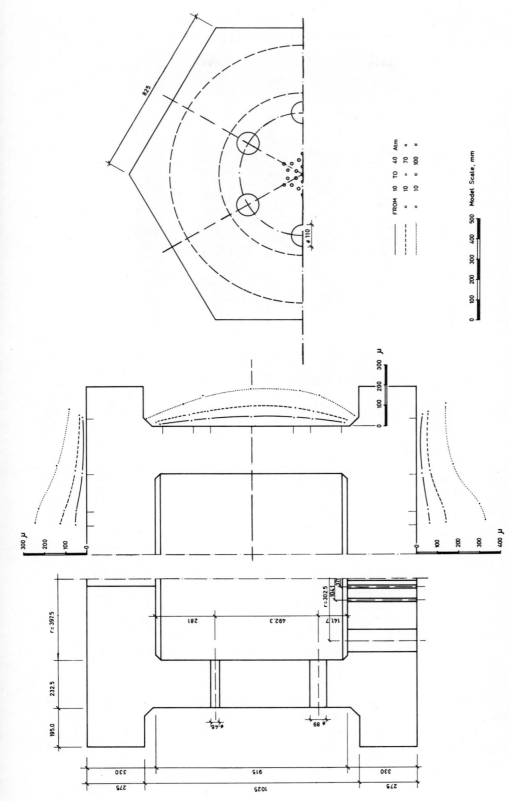

Fig. 69. Model of the THTR reactor pressure vessel. Diagram of the deformations obtained during tests with internal hydrostatic pressure

Fig. 70. Model of the THTR reactor pressure vessel. Cracking pattern found during the test to failure

The model tests on the whole structure showed that the end plates, contrary to the indications of an elastic analysis, had a margin of safety notably greater than that of the cylindrical body in the failure stage.

This evidence led to further tests to failure on a series of models of end plates with varying dimensions and reinforcement at a scale $\lambda = 20$. Fig. 71 shows the failure modes of five different end plates after tests to failure.

5.2.7 Experiments Carried out at the Cement and Concrete Association Laboratories

Among the most experienced research laboratories in the field of reinforced and prestressed micro-concrete models must be included that of the Cement and Concrete Association at Wexham Springs, England.

The following brief descriptions illustrate significant examples of their work.

a) Model of a Bridge Section for the Western Avenue Extension

The model reproduced in microconcrete at a scale of 1:16 a typical span of the Western Avenue Extension of length 63.5 m, width 26 m and depth 2.8 m [37].

The structure consisted of three box sections with two lateral cantilevers (Fig. 72) prestressed longitudinally, transversely and, in the vicinity of the supports, vertically also (Fig. 73).

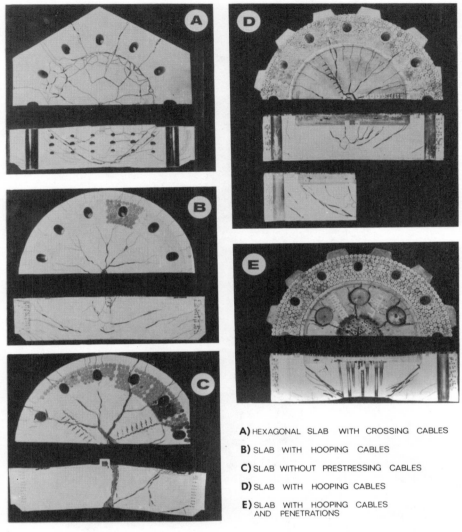

A) HEXAGONAL SLAB WITH CROSSING CABLES

B) SLAB WITH HOOPING CABLES

C) SLAB WITHOUT PRESTRESSING CABLES

D) SLAB WITH HOOPING CABLES

E) SLAB WITH HOOPING CABLES
AND PENETRATIONS

Fig. 71. Models of different end plates for the THTR reactor pressure vessel

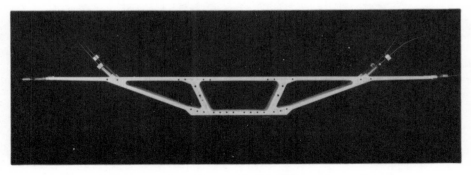

Fig. 72. Transverse section of the model of the Western Avenue Extension

Fig. 73. Model of the Western Avenue Extension. General view of the model during testing

Fig. 74. Model of the Western Avenue Extension after the final test to failure

The effect of prestressing and the stress distribution due to both dead and live loads was investigated by means of strain and displacement measurements. Finally the model was tested to failure under the combined action of bending, torsional and shear loads.

Fig. 74 shows the model after the test to failure.

b) Model of the Transverse Section of the West Gate Bridge, Melbourne

The design consisted of a central box structure to the sides of which were fixed, by means of transverse prestressing cables, the supporting cantilevers of the lateral decking [38].

The model reproduced one of the lateral cantilevers with its associated connection with the central beam.

During the test cycles the behaviour of the cantilever was progressively examined as the transverse prestress was applied. The efficiency of the upper and

Fig. 75. Transverse section of the West Gate Bridge, Melbourne (Australia). General view of the model during testing

lower connections was also checked. The behaviour of the structure under the effect of the dead and live loads was then determined (Fig. 75).

Finally the test was taken to the limit up to the collapse of the structure. Fig. 76 shows a detail of the lower end of the strut after the test to failure. In relation to this point, the stresses were firmly held under control during the whole course of the tests.

Fig. 76. West Gate Bridge. Detail of the lower end of the strut after the test to failure

c) Model of the Roof of Smithfield Market (Fig. 77)

The model, to a scale 1 : 12, was of an elliptic paraboloid reinforced concrete roof of plan area 68.5×39 m and thickness 75 mm supported at the boundary on columns at intervals of 7.6 m [15].

The behaviour of the structure when subjected to uniformly distributed symmetric and assymetric loads was measured by means of strain and displacement transducers. In addition the effect of prestressing the edge beams was investigated.

In Fig. 78 appears a view of the model after the final test to failure, showing a predominantly shearing mode of failure.

Fig. 77. Model of the roof of Smithfield Market during testing

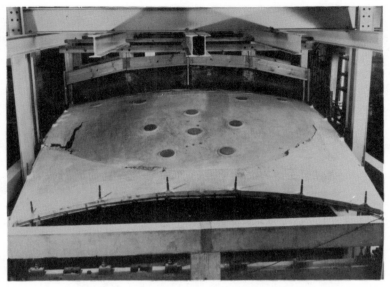

Fig. 78. Model of the roof of Smithfield Market after the test to failure

d) Model of the Metropolitan Cathedral, Liverpool

This structure is circular in plan having a diameter of 105 m (at the foundation level) and height 67 m [40].

The load bearing structure consists of 16 vertical reinforced concrete ribs (Fig. 79) connected by horizontal ring beams (also of reinforced concrete) at three separate levels.

Fig. 79. Model of the Metropolitan Cathedral of Liverpool during construction

The spaces between the ribs are covered with prefabricated concrete panels.

The behaviour of the structure was investigated on a micro-concrete model, reproduced at a scale of 1:28.9, subjected to the following load systems:

gravity load;

wind load;

effects of shrinkage of the ring beams and panels on the behaviour of the vertical ribs;

final test to failure, simulating the load due to impact of an aeroplane at the level of the central ring beam (Fig. 80).

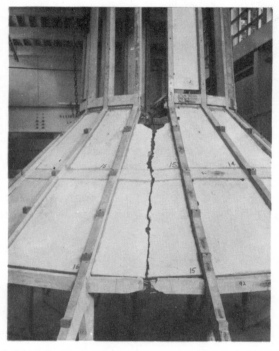

Fig. 80. Model of the Metropolitan Cathedral of Liverpool after the test to failure

e) Model of a Hyperboloid Cooling Tower

The model reproduced, at a scale of 1:25, a hyperboloid cooling tower of height 100 m and maximum diameter at the base 90 m; the thickness was approximately 125 mm and augmented locally at the upper and lower extremities.

The research was intended to investigate the statics of the tower after the collapse during a gale of some similar towers due to instability following the predominantly membranal behaviour of the structure.

The model was subjected to the action of self weight load applied by means of prestressing cables acting along the straight line generators of the hyperboloid (Fig. 81).

Fig. 81. General view of the model of a hyperboloid shell cooling tower undergoing test

Fig. 82. Detail of the model of a hyperboloid shell cooling tower after failure

The stability of the structure was then investigated when subjected to wind loading (applied by means of distribution plates and steel load rods) and to the effect of concentrated forces with particular reference to the local effect of the support legs.

Finally the structure was tested to failure with combined wind and self-weight loading.

Fig. 82 shows a detail of the structure after the collapse.

5.3   Model Experiments with Ratios of Efficiency $\zeta > 1$

5.3.1 Introduction

This type of the modelling is normally resorted to when, for some reason, we wish to investigate the behaviour of the entire structure up to failure but the difficulty and encumberance of providing a suitable self weight loading system makes advisable or enforces a reduction in the intensity of the required loads.

The following sections give details of typical cases of models of this type.

5.3.2 Model of a Small Aqueduct

The model was associated with the design of a prestressed aqueduct for irrigation, 7 m long and supported on saddles at its ends (Fig. 83) [11] [23].

To facilitate the hydrostatic loading tests, at a scale $\rho = 1$, the model was constructed from a pumice and cement mortar with ratios of similitude $\zeta = \lambda = 3$.

The tests at normal load were executed using a bag of water; the tests at greater loads with heavy liquids providing up to three times the normal maximum load required for approval.

Adjustment and control of the prestress in the ungrouted cables was achieved with spring dynamometers positioned at one end of the aqueduct.

5.3.3 Model of a Shed for Storage of Clinker

This consisted of walls for containing the thrust of stored clinker and of main pillars on lines 9 m apart for supporting the roof, acting in part as supporting ribs for the containing walls as well as for the roof itself.

The roof consisted of a corrugated shell supported laterally by tension chains passing between small pillars fixed at intervals of 3 m. These supported the shell on the crane beam underneath (Fig. 84). This beam was itself supported on the main pillars referred to above [11] [23].

In this case the scales of similitude chosen were $\zeta = \lambda = 8$. The model was constructed from a mortar of cement filled with magnetite sand (density 2.5 ton/m³) to achieve the scale ratio $\rho = 1$.

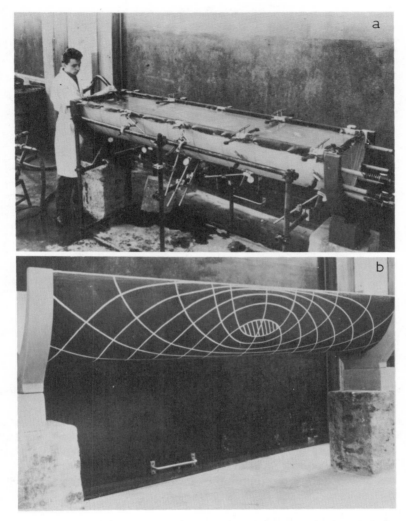

Fig. 83. a) General view of the model of an aqueduct in prestressed concrete during testing,
b) showing the isostatic lines of stresses

The clinker was sieved so as to produce a grading curve similar to the original but at a scale $\lambda = 8$.

The tests consisted of the application of:

the thrust of the clinker

the load relative to the self weight of the crane and its carrying capacity

live load on the shell roof, this latter applied by means of small bags of sand placed in isolated heaps to avoid continuity.

The final test to failure was carried out by incrementing the live load on the shell to a value corresponding to 900 kg/m² in the prototype. These tests were performed in the presence of the pressure of the clinker acting against the walls.

Fig. 84. Model of a shed for storing clinker

The cracking and collapse of the shell roof was investigated for this loading system.

As an illustration Fig. 85 shows the deflections and stresses in the containing walls and the main pillars due to the thrust of the clinker, and also the deflections of the shell roof both for a load distributed uniformly over the whole roof, and load limited to the region between the crown and one springer.

It was also possible to monitor the tensions in the linking chains with the following results:

270 kg/cm^2 for the thrust of the clinker
460 kg/cm^2 for the self weight of the roof
$\underline{310 \text{ kg/cm}^2}$ for a live load on the roof $= 150 \text{ kg/cm}^2$
$1{,}040 \text{ kg/cm}^2$ total specific force.

The influence line along the crane beam was determined for displacement of the crane along the track.

5.3.4 Model of the Pirelli Skyscraper

The building, in reinforced concrete, is 135 m high of which 125 m is above ground and constructed on 35 floors each 70 m long by 19 m wide [11] [23]. It is principally constructed on four supporting structures: the two inner ones

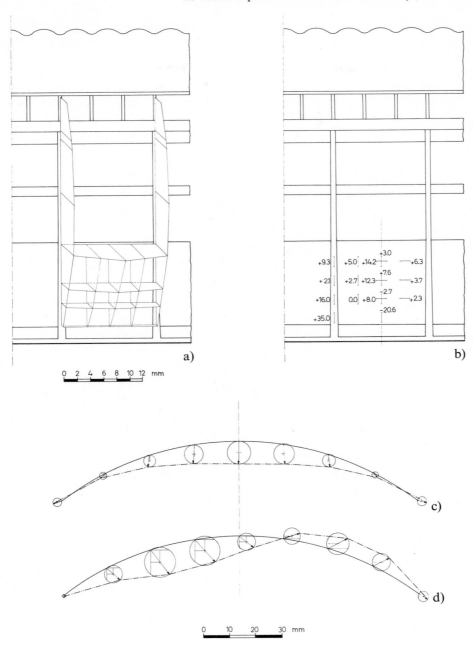

Fig. 85. Stresses and displacements obtained from the tests on the model of a shed for storing clinker: a) displacements of the main pillars and of the containing walls, b) stresses in the pillars and containing walls, c) displacements of the shell roof due to a load uniformly distributed over the whole shell, d) displacement of the shell roof due to a load uniformly distributed over half the shell

consist of pairs of pillars of trapezoidal section with minor sides opposed and the two end ones, again represented by opposed trapeziums and completed by vertical walls to form closed box structures. On one side, in the centre of the structure, are erected the lift shafts.

The model (Fig. 86) was reproduced with scale ratios $\zeta = \lambda = 15$. To reduce the encumbrance within the model alternate floors only were reproduced, each of these being assigned twice the normal moment of resistance.

Fig. 86. General view of the model of the Pirelli skyscraper (Milan)

Additional reinforcing was placed in the ceilings to avoid failure there since, as already discussed in section 5.2.1, the ceiling structure was investigated on a separate model.

For the practical need to avoid over encumbering the model other devices were used in the schematisation that were not completely in accordance with the statics of the structure.

Up to the level 38 m above ground internal connecting architraves were reproduced between corresponding pairs of pillars at each floor level. Subsequently, after examining the behaviour of the structural model, it was decided to complete the installation of the architraves up to the summit to augment the stiffness of the model.

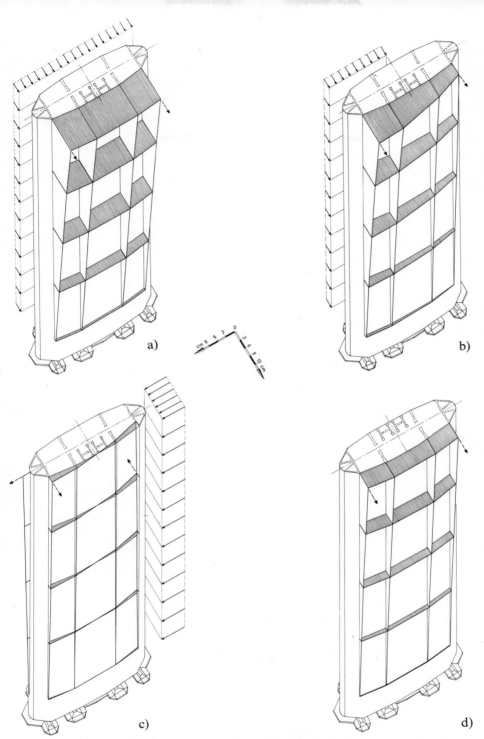

Fig. 87. Model of the Pirelli skycraper. Deflection diagrams for different load conditions:
a)
b)} wind loading (intensity 112 kg/m²),   c) wind loading (intensity 112 kg/m²),
d) dead load on the pillars and floors and live load on the floors

8   Fumagalli, Statical Models

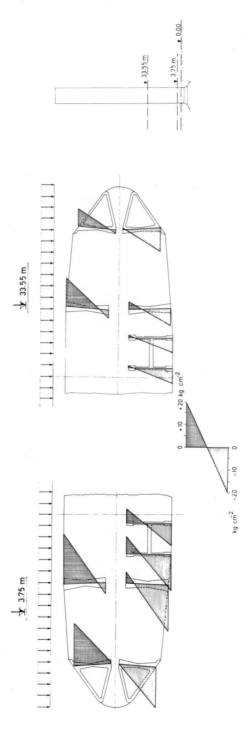

Fig. 88. Model of the Pirelli skycraper. Stress distribution in the pillars due to wind loading (symmetrical load of intensity 112 kg/m²)

The loading system for both dead and live loads consisted of rubber ring dynamometers distributed on each floor as well as (in the case of self weight only) to the pillars and vertical partition walls. The rings were taken back to a tubular structure and loaded by means of hydraulic jacks at the foot of the model.

Wind loading was also applied using ring dynamometers arranged in three separate systems. The first for the thrust normal to the minor axis applied to one end, the second for the longitudinal vertical wall and finally to half the surface included between the central axis and one end.

The intensity of the wind load assumed in the design calculations was $112\,kg/m^2$ uniformly distributed over the surface.

Among the points of interest that emerged from the subsequent tests was the eccentric stiffening induced by the lift shafts, indicated by the lateral deflection produced by the self weight load. This effect is not detected in the real structure since it is corrected automatically during construction, but at the same time it induces a non-uniform stress distribution in the pillars.

This structural eccentricity produces a limited torsional deformation effect for wind loading acting normal to the minor axis (Fig. 87).

This suggests a modified structural tendency of partial constraint of the lift shafts by the rest of the structure (Fig. 88).

It is of interest to examine the stress distribution in the pillars due to wind effect and in particular to separate the bending action of the pillars acting independently from the combined supporting action produced by the constraint of the interconnecting architraves.

After completing the interconnecting architraves between the pairs of pillars from level 38 m to the summit the maximum deflection normal to the major axis for a wind loading of 112 kg/m² was reduced from 17 cm to 11.5 cm (Fig. 87).

By suddenly removing the wind load it was possible to measure the free fundamental resonant frequency of the model and from this, by a simple extrapolation, the corresponding value for the prototype. For flexure normal to the major axis this gave a period $t=3.80$ secs and for torsional action $t=2.40$ secs.

Subsequent research on a model tested in a wind tunnel permitted the determination of the wind loading patterns for varying angles of incidence. From these tests it was observed that the aerodynamic behaviour of the structure was not greatly dissimilar to that of an aeroplane wing.

References

[1] Alami, Z. Y., Ferguson, P. M.: Accuracy of Models Used in Research on Reinforced Concrete. ACI Journal, Proceedings, Vol. 60, Nr. 11, 1643—1661 (November 1963).
[2] Alexander, Benedict, F: Use of Micro-concrete Models to Predict Flexural Behavior of Reinforced Concrete Structures under Static Loads. Research Report R 65-04, School of Engineering, Massachussetts Institute of Technology (March 1965).
[3] Amaratunga, M.: Tests on Models of Concrete Structures. Engineering, London, Vol. 194, Nr. 5021, 62—63 (July 1962).

[4] Best, B. C.: Testing Micro-concrete Structural Models. Presented to the Joint British Committee for Stress Analysis ad a Meeting entitled: Model Testing Techniques, the Collection and Interpretation of Data. University College, London (June 1967).

[5] Billington, D. P.: Thin Shell Concrete Structures. New York: McGraw-Hill. 1965.

[6] Borges, J. F., Lima, J. A. E.: Crack and Deformation Similitude in Reinforced Concrete. Rilem Bulletin, Paris, New Series Nr. 7, 79—90 (July 1960).

[7] Bouma, A. L.: Investigations on Models of Eleven Cylindrical Shell Made of Reinforced and Prestressed Concrete. Proceedings, Symposium on Shell Research, 79—101. Delft (1961). New York: Wiley-Interscience, and Amsterdam: North-Holland Publishing Company. 1962.

[8] Brading, K. F., McKillen, R. R., Finigan, A.: Elastic and Ultimate Pressure Tests on a One Tenth Scale Model of the Dungeness B Concrete Pressure Vessel. Proceeding of the Conference organized by the British Nuclear Energy Society, London (July 1969).

[9] Breen, J. E.: Fabrication and Tests of Structural Models. Proceedings ASCE, Vol. 94, Nr. ST6, 1339—1352 (June 1968).

[10] Davidson, I., Purdie, A. C.: Small Scale Model Prestressed Concrete Pressure Tests at Foulness. Proceeding of the Conference organized by the British Nuclear Energy Society, London (July 1969).

[11] Fumagalli, E.: The Use of Models of Reinforced Concrete Structures. Magazine of Concrete Research, Vol. 12, Nr. 35, 63—72 (July 1960).

[12] Fumagalli, E., Verdelli, G.: Static Tests of a Model of a Prestressed Concrete Pressure Vessel for a THTR Nuclear Reactor. Meeting concerning Experimental Investigation and Safety Aspects of PCRV's, Delft (December 1970).

[13] Harris, H. G., Sabnis, G. M., White, R. N.: Reinforcement for Small Scale Direct Models of Concrete Structures. Symposium on Structural Models, SP-24, American Concrete Institute, Detroit (1970).

[14] Hornby, I. W.: The Behavior of the Olbury Model Vessel with Time under Thermal and Pressure Loadings. Proceeding of the Conference organized by the British Nuclear Energy Society in London (July 1969).

[15] Jones, L. L., Base, G. D.: Tests on a 1/12th Scale Model for the Dome Shell Roof for Smithfield Poultry Market. Proceedings on the Institution of Civil Engineers, Vol. 30 (January 1965).

[16] Langan, D.: Model Philosophy in Relation to Prestressed Concrete Pressure Vessel Design Problems. Proceeding of the Conference organized by the British Nuclear Energy Society in London (July 1969).

[17] Lauletta, E.: Statics of Hyperbolics Paraboloidical Shells Studies by Means of Models. Proceeding of the Symposium on Shell Research, Delft (August-September 1961).

[18] Launay, P.: The Two Bugey 1/5 Models. Proceeding of the Conference organized by the British Nuclear Energy Society in London (July 1969).

[19] Lee, S. T.: Small Models of a Multiplanel Flat Plate Structure. M. S. Thesis, Massachussetts Institute of Technology, Cambridge, Mass. (February 1964).

[20] Little, W. A.: Paparoni, M.: Size Effect in Small-scale Models of Reinforced Concrete Beams. ACI Journal, Proceedings, Vol. 63, Nr. 11 (November 1966).

[21] Menon, S. K.: The Scandinavian Prestressed Concrete Pressure Vessel Model Project. Proceeding of the Conference organized by the British Nuclear Energy Society in London (July 1969).

[22] Neville, A. M.: A General Relation for Strenghts of Concrete Specimens of Different Shapes and Sizes. ACI Journal, Proceedings, Vol. 63, Nr. 10 (October 1966).

[23] Oberti, G.: Large Scale Model Testing of Structures outside the Elastic Limit. Bulletin ISMES, Nr. 12 (April 1959).

[24] Oberti, G.: La ricerca sperimentale su modelli strutturali e la Ismes. L'Industria Italiana del Cemento, Anno XXXIII, Nr. 5 (May 1963); Bulletin Ismes, Nr. 22 (January 1964).

[25] Oberti, G.: Model Analysis for Structural Safety and Optimization. Bulletin Ismes, Nr. 41 (February 1970).

[26] Preece, B. W., Davies, J. D.: Models for Structural Concrete. London: C. R. Books Limited. 1964.

[27] Rashid, Y. R., Ople, F. S., Chang, T. Y.: Comparison of Experimental Results with Response Analysis for a Model of a Pressure Vessel. Proceeding of the Conference organized by the British Nuclear Energy Society in London (July 1969).

[28] Rocha, M.: Model Tests in Portugal, Part 1 and 2. Civil Engineering and Public Works Review, London, Vol. 53, Nr. 619, 49—53 (January 1958); Nr. 620, 179—182 (February 1958).

[29] Rowe, R. E.: Tests on Four Types of Hyperbolic Shell. Proceedings Rilem, Symposium on Shell Research, Delft (1961). Amsterdam: North-Holland Publishing Co., and New York: Interscience. 1961.

[30] Rowe, R. E., Base, G. D.: Model Analysis and Testing as a Design Tool. Proceedings, Institution of Civil Engineers, London, Vol. 33, 183—199 (1966).

[31] Rowe, R. E., Best, B. C.: The Use of Model Analysis and Testing in Bridge Design. Preliminary Publication, 7th Congress of the International Association for Bridge and Structural Engineering, 115—121, Rio de Janeiro (August 1964).

[32] Scotto, F.: Techniques for Rupture Testing of Prestressed Concrete Vessel Models. Proceeding of the Conference organized by the British Nuclear Energy Society in London (July 1969).

[33] Scotto, F.: Thin-Walled 1:20 Prestressed Concrete Pressure Vessel Model for THTR Reactor Type. 1st International Conference of Structural Mechanics in Reactor Technology, Berlin (September 1971).

[34] Smith, J. R.: Problems in Assessing the Correlation between the Observed and Predicted Behavior of Models. Proceeding of the Conference organized by the British Nuclear Energy Society in London (July 1969).

[35] Somerville, G., Roll, F., Caldwell, J. A. D.: Tests on a One-twelfth Scale Model of the Mancunian Way. Technical Report TRA 394, Cement and Concrete Association, London (December 1965).

[36] Stefanou, G. D., Gill, S.: An experimental Investigation into the Behavior of Perforated End Slabs for Concrete Pressure Vessels under Temperature and External Load. Proceeding of the Conference organized by the British Nuclear Energy Society in London (July 1969).

[37] Swan, R. A.: The Construction and Testing of a 1/16th Scale Model for the Superstructure of Section 5, Western Avenue Extension. Cement and Concrete Association, Technical Report TRA 441 (July 1970).

[38] Taylor, H. P. J., Clements, S. W.: Test on a 1/9 Scale Model of a Transverse Section of the West Gate Bridge, Melbourne. Cement and Concrete Association, Technical Report TRA 425 (June 1969).

[39] White, R. N., Sabnis, G. M.: A Bibliography on Structural Model Analysis (with Particular Emphasis on Concrete Structure). Report Nr. 325, Department of Structural Engineering, Cornell University (September 1966).

[40] Model Testing. Proceedings of a One-day Meeting held in London on 17 March 1964, Cement and Concrete Association, London (1964).

6. Construction and Testing of Models of Concrete Dams in the Elasto-Plasto-Viscous Range up to Failure

6.1 Historical Background

Static model simulation of concrete dams, particularly arch dams, forms a field of research with wide and important applications and one in which the techniques of reproduction and testing have undergone successive stages of refinement and evolution, the result of much study and experience.

Starting in the years following the first world war, when the analytical design methods for arch dams were inadequate, experiments on models, even if limited to studies in the elastic range, underwent their initial and important development.

The years following the second world war saw a greatly increased interest in such models as a result of an increase in the number and importance of projects under investigation and following the establishment of some permanent research laboratories specialising in this field.

Following the development of methods of mathematical design analysis, experiments on models provided a basis of comparison for the validity of the analytical results. In the meantime through the refinement of modelling techniques experiments on models continued to be developed to the present advanced stage.

Models have also been used to examine the behaviour of the dam structure up to failure. In this case the rock abutments are required to provide the appropriate strength and stiffness to the constrained surfaces.

Finally during the 1960's with the development of geomechanical models, to be discussed in detail in Chapter 7, the most ambitious objective of examining the equilibrium of the whole system comprising the dam and associated rock masses exposed to the thrust of the reservoir, was tackled with notable success.

The choice and preparation of materials and suitable loading methods have already been discussed in detail in Chapters 2 and 3.

Returning to the present subject we will start by outlining the procedures of two of the laboratories that particularly specialise in this field.

6.2   L.N.E.C. Techniques

It has been pointed out that plaster-diatomite mixtures (2.3.2) are affected during curing by dehydration processes that induce heterogenities in the materials and associated internal stresses due to shrinkage.

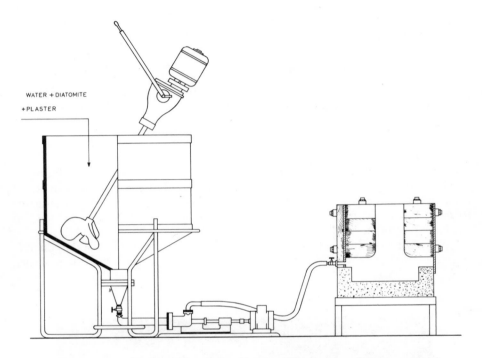

WATER + DIATOMITE

+PLASTER

Fig. 89. Diagram of the equipment for the production of plaster-diatomite material

At the L.N.E.C. in Lisbon the models are prepared in the following manner in order to avoid to a large extent these inconveniences [2]. The diatomite and the water are initially put into the mixer shown diagrammatically in Fig. 89, the plaster being added some hours later so as to reduce the formation of air bubbles. From the mixer the fluid material is transferred by a suitable pump into a rugged mould to form a block that roughly reproduces the system of the dam on its foundation. The thickness of the body of the dam in the untreated block is at least three times greater than that of the finished model (Fig. 90).

At the same time test samples are prepared to monitor the properties of the material.

Both model and samples are then put aside to cure and dehydrate for a period of one to two months in conditions of 40% relative humidity and 35° C. This is to ensure slow and homogeneous curing, particularly in the centre of the mass, and thus greatly reduce the shrinkage and internal stresses.

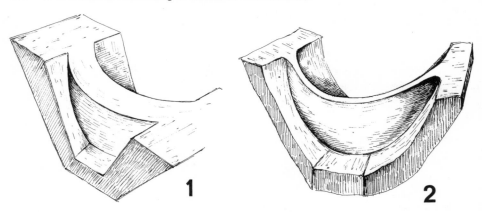

Fig. 90. Preparation of castings for a plaster-diatomite model: (1) roughly cast block, (2) finished model

In the case of arch dams the model is shaped with a fraise rotating at 50,000 r.p.m. mounted on a movable radial arm. This latter is adjustable in height on a vertical shaft fixed at the centre of the radius of curvature of the arch being formed. For other types of dam (Fig. 91) the final machining is carried out with a pantograph.

When the modulus of elasticity of the rock is of the order of $^1/_{10}$ of the concrete recourse is made to casts of different properties for the dam and for the different rock

Fig. 91. Example of a model finished by machining with a pantograph

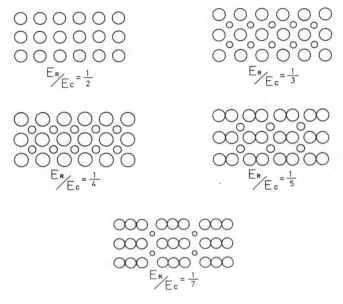

Fig. 92. Illustration of various perforation systems for reducing the moduli in different ratii E_r/E_c

zones, adding to the mix relative to these latter granules of polystyrene and barytes in suitable proportions. For greater modular ratios holes are made in the model rock according to the schemes shown in Figs. 92, 93, and 94.

Fig. 93. Details of support for ratio $E_r/E_c = 1/5$

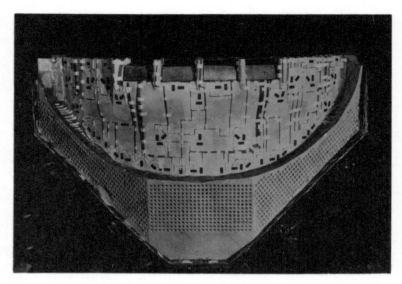

Fig. 94. Model showing application of electrical resistance strain gauges and ratio $E_r/E_c = 1/2$

In the case of Fig. 95 faults have been reproduced passing right through the model abutments.

The scales of the models for the most part lie between 100 and 500.

The tests are normally carried out in duplicate on two models in parallel as a safety check on the measurements.

Fig. 95. Reproduction of a model with faults

The models are used principally for measurements in the elastic range under the action of hydrostatic loads, loading being normally on the upstream face through bags of mercury.

Strain measurements are carried out with electrical resistance strain gauges.

When the models are tested to failure (Fig. 96) loading is normally by means of inclined jacks, as discussed in section 3.5.

It is clear that in the models described above the tests to failure are intended to investigate the behaviour of the dam wall when constrained elastically by its foundation.

Fig. 96. Model tested to failure

6.3 Ismes Techniques — Conventional Models[7]

Experiments on large models make possible a full and detailed investigation of the structure and do not represent in themselves an excessive cost if we consider that the work of constructing the model and setting up the loading and measuring systems does not normally account for more than 50% of the total cost.

The models are reproduced at a reduced scale varying between $30 < \lambda < 120$ depending on the size and importance of the project [8].

The model is positioned in a rigid containing structure of highly reinforced concrete of suitable size for the dimensions of the model (together with a sufficiently extensive reproduction of the associated rock foundation) and the type of test to be conducted.

---

7 The description "conventional" is used to distinguish the static models described below from geomechanical models which are dealt with in Chapter 7.

Fig. 97. Installation of loading rods for self weight

After setting out the outline of the model the loading rods for the self weight loads are positioned (Fig. 97) and then the model is cast using a pumice and cement mixture that suitably represents the particular rock type with properties reduced in the preselected scale ζ, thus giving the modular ratio between rock and concrete decided upon at the planning stage of the model.

In the preliminary stage of the investigation suitable mixes are designed to reproduce both the concrete of the dam and the surrounding rock mass and the properties of these mixes established from preliminary tests on samples. At the same time a preliminary wooden model of the dam model is constructed to the same scale to a tolerance of ± 1 mm (Fig. 98).

Following the necessary review and possible revision of design detail, the preliminary model is used to prepare plaster moulds of the surfaces of the dam faces and joints, for casting the experimental model.

Finally the preliminary model provides an admirable base on which to draw the isostatic lines obtained from the experimental stress analysis (Fig. 99).

After fixing the moulds in position alternate blocks of the experimental model are cast, reproducing in general $^1/_2$ to $^1/_3$ of the total number of radial joints envisaged in the prototype (Fig. 100). When the mould is dismantled the model is coated with impermeable varnish (vinyl or epoxy resin) and then with layer of protective varnish to ensure a uniform state of humidity, as discussed previously as the technique of "humid models" (see 2.3.3).

Fig. 98. Preliminary wooden model

Fig. 99. Isostatic lines traced on a model (downstream face)

About 20 days after casting, the radial joints between blocks are injected with a three part liquid mix (cement, water and bentonite) to a pressure of 0.2—0.3 kg/cm² to ensure continuity of contact without appreciable cohesion.

In the meantime the loading systems for self weight and hydrostatic loads are set up and by about 30 days after casting the model is considered to be sufficiently matured and ready for testing.

Fig. 100. Cast of alternate blocks of a model

Fig. 101. Model of "El Novillo" dam (Mexico) during testing showing use of Huggenberger
mechanical extensometers

In order to control the properties of the materials the model is gradually loaded up to a predetermined level through a series of loading and unloading cycles. These load cycles absorb the initial jack friction and the small initial anelastic deformations. The model is then ready for the test cycles in the elastic range. These latter are carried out between a lower limit γ_0' and γ_1', where γ_0' is normally about 10—15% of γ_1' and is required for the consolidation of the model structure.

Strain measurements are normally made on the downstream face with Huggenberger mechanical extensometers (see 3.7) with gauge lengths of 50—75—100 mm (Fig. 101) and on the upstream face with Galileo electro-acoustic extensometers of 40—60 mm gauge lengths reading out to a remote central terminal. Strain measurements are normally carried out on a rosette in four directions. Overall deformations are determined using dial gauges reading to 0.01 mm.

Fig. 102. View of the model of "Hongrin" dam (Switzerland) during the test to failure

Recently there has been a tendency to replace the extensometers and dial gauges indicated above by Hottinger type inductance instruments with centralised automatic digital read out.

On completion of the tests in the elastic range subsequent tests are carried out with the load increasing up to failure (Fig. 102).

In this final stage the deformation of the most important points are measured with electrical transducers linked to a data logger.

6.3.1 Objectives and Results of the Investigations

From the experiments we normally obtain, after the necessary examination and analysis of the experimental data, the distribution of the four non-zero stress components on the two faces together with the directions and magnitudes of the principal stresses.

The displacement measurements define the deformed surface of the downstream face of the dam showing, in general, at each measuring point the three basic components; normal to the cantilever, tangential to the cantilever and tangential to the arch.

Other control measurements are normally made across the joints or at the base of the principal cantilever to determine the eventual separation movements.

The more advanced mathematical methods of analysis now allow us to obtain results sufficiently coincident with the experimental results obtained from models, provided that the dam is of sufficiently regular shape and relatively symmetric, particularly with respect to the boundary constraint conditions[8].

Nevertheless design analyses based on models are generally more frequently requested, especially when the model can give a more complete and detailed picture of the information of interest.

This is particularly so when interest is focussed on disturbances associated with irregularities of the boundary constraints, whether these irregularities are due to variations in the rock properties or to a random development of orographic movement of the rock mass; as likewise when one wants to establish the influence of known discontinuities such as perimetral, radial or sub-vertical joints, or of openings or spillways forming intermittent local weakening or stiffening structures, all these being features that can substantially alter the regular stress pattern.

There are also cases in which new shapes or details of design are unfavourable to classical methods of mathematical analysis.

In all these cases only models, with their ability to comprehensively and faithfully reproduce the most complex statical behaviour of the structure, can produce the most reliable statical analysis.

6.3.2 Criteria for Determining the Distribution of Self Weight Load

Self weight tests carried out on models with closed joints, thus presenting a continuous arch, represent a limiting case not entirely realised in reality. The possibility that we investigate the other limiting hypothesis, that of a purely gravitational distribution (imagining the entire structure to be constructed of independent blocks with joints open at all levels), is normally excluded due to the possibility of instability occurring in the free standing cantilever, particularly in the crown region when the overhangs are not negligible.

---

8 There are now several cases of comparisons that show good agreement between results obtained from theoretical analyses and from models ("Tachien" dam, Formosa — "Chiotas" dam, Italy — "Emosson" dam, Switzerland — "Tirso" dam, Italy).

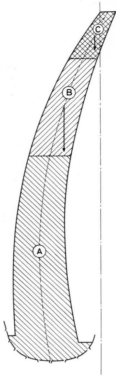

Fig. 103. Investigation by zones for the self weight load according to the planned construction programme

For experimental purposes one can consider a third hypothesis, intermediate between the two previous ones, more in keeping with the presumable constructional requirements noted previously in section 3.2.1.

In this case the cantilever is ideally divided into three zones (Fig. 103):

A) lower zone, with open joints, that can maintain itself in free standing equilibrium,

B) central zone that can only maintain itself in free standing equilibrium on the hypothesis of a closing of the joints in zone A,

C) upper zone, in equilibrium only on the hypothesis of closing of the joints in zones A and B.

For zone A an analytical treatment assuming purely gravitational loading is sufficient.

The self weight load for zone B is applied to the model after the radial joints in zone A have been injected; the measurements giving a picture of the stress distribution transmitted from zone B to zone A.

In this context it is worth noting that even in the part of zone A overhanging upstream the injection process can, without firmly bonding the blocks of the dam, induce constraints due to the resulting continuity that, if only irregularly distributed, permit the transmission of traction forces of the magnitude of several kg/cm^2 across the surfaces of the joints.

In the same way the loads relative to zone C are applied after the joints of zones A and B have been completed injected. The measurements show for the record the picture of the distribution of stresses transmitted from zone C to the zones A and B.

By superposition one finally obtains a picture of the total resulting stress distribution that more closely reproduces the statical situation resulting from the on site construction programme.

Indeed the blocks overhanging downstream in the crown zone, on becoming plastically deformed, frequently come into contact with each other along the joint surfaces, producing a structural continuity even before injection. Also, during the first cycles of material placement and the first cycles of heat generation during curing, and also following any differential adjustment effects, a part of the load that was initially distributed gravitationally can become successively transferred to a spatial distribution by continuous arching. For this reason the two situations investigated, that of continuous arching and that of distribution in successive stages, provide the two limiting conditions between which, to a good approximation, one can assume that the more complex real situation will lie.

6.3.3 Distribution of the Hydrostatic Load

As in other areas experiments on large models facilitate the analysis of details of the stress and deformation distributions.

Fig. 104.   Static model of "Almendra" dam (Spain) after the experiment to failure

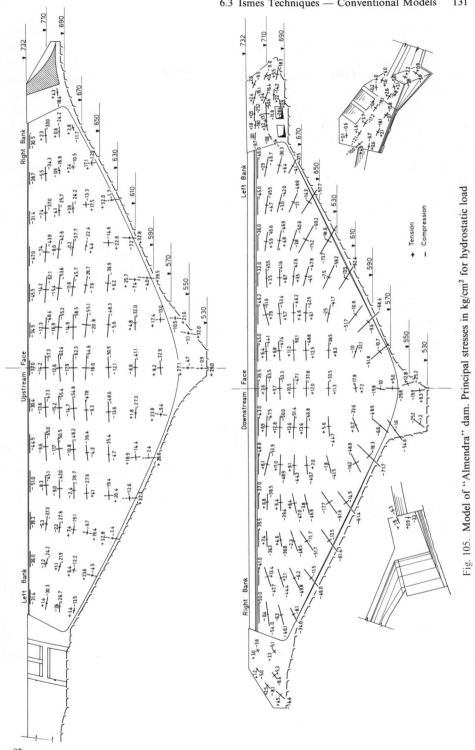

Fig. 105. Model of "Almendra" dam. Principal stresses in kg/cm² for hydrostatic load

They make possible refined analyses and ellucidations of the structural behaviour that allow one, among other things, to reconstruct as a system of discrete
approximations the individual resistant force components forming the stress
distribution due to the hydrostatic load. As an illustration, we will consider the
static model of the "Almendra" dam[9] (Fig. 104) reproduced at a scale $\lambda = 100$.
Knowing the stress distribution due to the hydrostatic load on the two faces (in
this specific case the principal stresses are shown directly [Fig. 105]), the analytical
procedure starts from the assumption that the structure behaves elastically as a
shell and the stresses vary linearly through its thickness (h).

Referring to a set of orthogonal axes with the x axis tangential to the
horizontal arch and the y axis tangential to the sub-vertical cantilever the force
components (stress resultants) acting on an elementary block are:

Membrane Effect: Membrane stress resultants

N_x = Normal membrane force per unit length in the arch direction.
N_y = Normal membrane force per unit length in the cantilever direction.
T_{xy} = Membrane shearing force.

Bending Effect: Bending stress resultants

M_x = Bending moment per unit length in the arch direction.
M_y = Bending moment per unit length in the cantilever direction.
$M_{xy} = M_{yx}$ = Twisting moments per unit length.

$Q_x \quad = \dfrac{\partial M_x}{\partial x} =$ Shear force per unit length in the arch direction.

$Q_y \quad = \dfrac{\partial M_y}{\partial y} =$ Shear force per unit length in the cantilever direction.

Only the three membrane forces and the three bending moments are
independent.

The values of these components can be derived from the experimental results
by the following expressions:

Membrane Effect:

$$N_x \;=h\,\frac{\sigma_u+\sigma_d}{2}$$

$$N_y \;=h\,\frac{\sigma'_u+\sigma'_d}{2}$$

$$T_{xy} \;=h\,\frac{\tau_u+\tau_d}{2}$$

---

9 "Almendra" dam (Spain); height 202 m, chord 520 m, volume of concrete 2,200,000 m³
in the central shell of the dam.

Bending Effect:

$$M_x = \frac{h^2}{6} \frac{\sigma_u - \sigma_d}{2}$$

$$M_y = \frac{h^2}{6} \frac{\sigma'_u - \sigma'_d}{2}$$

$$M_t = M_{xy} = M_{yx} = \frac{h^2}{6} \frac{\tau_u - \tau_d}{2}$$

in which σ'_u and σ'_d are the vertical stress components on the upstream and downstream faces, σ_u and σ_d the corresponding horizontal components and τ_u and τ_d the shear components.

Letting P be the hydrostatic pressure at the middle surface, the partial differential equation of equilibrium takes the following form:

$$P = P_{am} + P_{cm} + P_{ab} + P_{cb} + P_t$$

where:
Membrane Contribution:

$$P_{am} = \frac{N_x}{r_h} \sin \phi, \text{ the axial arching resistance}$$

$$P_{cm} = \frac{N_y}{r_v}, \text{ the axial cantilever resistance}$$

Bending Contribution:

$$P_{ab} = -\frac{1}{r_h} \frac{\partial (M_x \cos \phi)}{\partial y} + \frac{\partial^2 M_x}{\partial x^2}, \text{ the arch bending resistance}$$

$$P_{cb} = \frac{1}{r_h} \frac{\partial^2 (M_y r_h)}{\partial y^2}, \text{ the cantilever bending resistance}$$

$$P_t = \frac{2}{r_h} \frac{\partial M_t}{\partial x} \frac{\partial r_h}{\partial y} + \frac{\partial^2 M_t}{\partial x \partial y} + \frac{1}{r_h} \frac{\partial^2 (M_t r_h)}{\partial x \partial y}, \text{ the twisting resistance}$$

r_h and r_v represent respectively the local radii of curvature in the horizontal and vertical planes and ϕ is the angle that N_y makes with the horizontal.

The results of such an analysis are shown in Fig. 106 for the central cantilever and some lateral cantilevers. For these latter mean stress values have been used from the symmetrical sections. The satisfactory symmetry of these values makes this procedure permissible.

For the central cantilever the arching contribution for the entire interval between level 692 m and the crest is greater than the total diagram of the applied hydrostatic load.

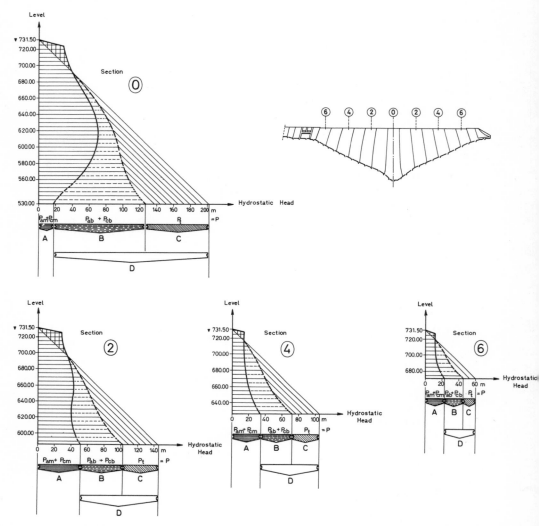

Fig. 106. Distribution of hydrostatic load in the crown section and in several symmetric lateral sections: (A) membrane resistance, (B) flexural resistance, (C) torsional resistance, (D) combined bending resistance

The excess load is in particular the result of the relatively high stiffness of the crest arch, on which the cantilever is effectively supported.

The membrane contribution reaches its maximum value around level 620 m, covering in this zone about 70% of the entire load diagram.

For the lower part of the dam, below level 620 m, the arch and cantilever bending contributions become consistent together with the torsional contributions, this being due to the limited aperture of the arches, of considerable thickness, that produces a predominantly bending action in this region.

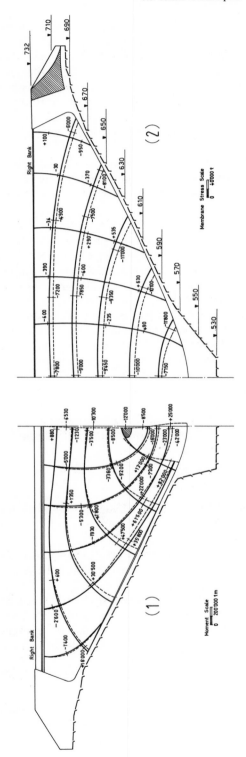

Fig. 107. (1) Isostatics for bending action and diagram of moment distributions on the same section,
(2) isostatics for membrane action and diagram of axial stresses on the same section

It is useful to note that moving from the crown section to the abutments the bending and torsional contributions become predominant by virtue of the reduced height of the cantilevers and the smaller membrane contribution. This reduction follows principally from the gradual reduction of the curvature of the arch on moving from the crown towards the abutments, always starting from the hydrostatic load stress distribution and assuming linear variation through the thickness. One can likewise extend the analysis to the force resultants acting in the whole shell, separating the contributions due to membrane action from those due to bending action.

We achieve, therefore, a subdivision of the stress field into two components that we call respectively the membrane effect and the bending effect, from which we obtain the principal stresses and families of isostatic curves relative to the membrane and bending components (Fig. 107).

From these analyses we note that:

for membrane action, the isostatics form a highly reticular pattern for both arches and cantilevers, and that only in the plug zone do the arch isostatics develop slowly "dipping" lines. Examining the diagram of forces that accompanies the isostatics it can be seen that the thickness of the design is well proportioned in relation to the arching resultants that are developed in the structure;

in the case of bending action the family of isostatics that start from the abutments develop along appreciably curved lines leading to a zone of uniform moment positioned approximately at level 620 m in the region of the central cantilever. The bending stresses in particular show an appreciable resisting contribution, tending to support the central body of the shell.

It is also worth noting the bending stresses relative to the isostatics normal and conjugate to the former ones along the line of the abutment. The positive moment developed between levels 750 and 650 m indicates a local rigidity and consequent attraction of load, through the contribution of its bending resistance, to support the greater flexibility of the zone above.

6.3.4 Limits of Applicability of Conventional Models for Investigating Geomechanical Type Failures

Coming now to consider the response to failure of the abutment rock system, reproduced according to the techniques discussed, it should be remembered that conventional models should in principle:

1. be considered reliable where failure occurs by bond failure or perimetral and superficial crushing along the surface of the abutment and where the average deformability and strength properties of the materials are respected.

It is worth noting that perimetral bond failure frequently precedes or accompanies the general collapse of the structural wall;

2. on the other hand be considered unreliable concerning failure of the rock and, by reflection, of the same structural wall from the instant of onset of failure at depth in the rock mass of the abutments. In substance this means that the execution of tests to failure on conventional models should be considered valid only on con-

dition that the stability of the rock mass is not under discussion. In fact, it should be remembered that in these models the specific weight contributed by the model rock is insignificant compared with that required by equation (10), against which the material reproducing the rock is cohesive, homogeneous and isotropic.

On the limited validity of such a model it is sufficient to consider that whilst the self weight stresses increase in depth with the thickness of overburden the cohesion remains constant for the whole system.

For these reasons conventional models cannot be used for an overall investigation to failure of the system formed by the dam with its associated abutment rock masses.

There are, however, particular cases in which conventional type models can assist the examination, in the right situation, of particular geomechanical problems. This happens for example in the case of local accidents or else when the object of the examination is an analysis limited to zones of weakness due to insufficient orographic development or reduced mechanical strength [12].

In this case it is sometimes necessary to extend the application of self-weight loads to part of the rock system as a fixed stabilised load. This is achieved at Ismes by an independent loading system using load rods and spring dynamometers.

On the other hand, it is not practically possible to apply this system to the entire mass as this would be disproportionately difficult and cumbersome.

In the case of the model of the "Hongrin" dam (Switzerland) the central rock spur, that bore the combined thrust of the abutments of the two adjacent arches, was stratified on planes inclined upstream at 30° (Fig. 102). In the model an angle of friction $\phi = 40°$ and a cohesion of 1.3 kg/cm² was assigned to the surfaces of contact between the strata, this latter value already reduced in the ratio $\zeta = 3.5$, to reproduce a limited bond existing between the strata.

In this case it was evident that the self weight loading of the spur formed a determining component for the general equilibrium against sliding.

A self weight loading system, limited to the potentially sliding mass (Fig. 97), of suitably distributed loading bars was applied to this spur.

During the test to failure, albeit at an amply sufficient margin of safety, sliding of the strata occurred starting from the left abutment of the more important arch, positioned on the right of the rock spur.

The movements of the abutment following the sliding of the strata produced a diffuse and extensive fissuration in the body of the concrete dam; this being the determining cause of its final collapse (Fig. 102).

This model provides a classical example of the possibility of carrying out a test to failure by increasing only the hydrostatic load, maintaining the normal value of the self weight effect.

If it is assumed, as a pure hypothesis, that the limiting equilibrium in respect of sliding (ignoring the contribution of cohesion) occurs at the onset of failure, it is evident that increasing proportionately the two loads should show, from this standpoint, the existence of a theoretically unlimited factor of safety against sliding.

On the other hand a small increase of the hydrostatic load should produce general sliding of the strata.

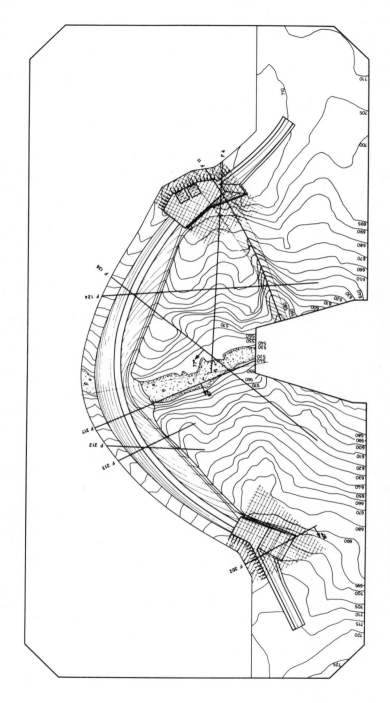

Fig. 108. Plan of "Almendra" dam showing positions of faults; rock abutment zones to which self weight loads were applied are shown shaded

As a general criterion it is advisable to carry out the test to failure by applying the least favourable load conditions to the model. In fact the two loads are normally increased up to 3 and sometimes 4 times the normal design loads, provided that deformations are more or less limited to the elastic range so as not to damage the model before the final test to failure.

In the case of faults that cross the rock mass with nearly vertical dips the stress flow due to self weight does not appreciably influence the equilibrium against sliding between the surfaces of the fault. Unsuccessful reproduction of the force of gravity is not, therefore, critical in this particular case.

This was the case for the already cited model of the "Almendra" dam (6.3.3), for which near vertical faults broke up the abutment rock masses, especially on the left bank, into independent blocks (Fig. 108).

The reproduction of such slip surfaces is important because of the static disturbances they can induce in the dam structure.

A geomechanical study that was capable of being developed using a conventional model is illustrated by the model of the "Mequinenza" gravity dam (Spain) ($\lambda=60$, $\xi=4$).

This was possible because, dealing with a gravity dam, the investigation was carried out on a plane model of limited size by applying a self weight loading system to both the dam section and the associated rock foundation using load bars with ring dynamometers for the dam and spring dynamometers for the rock (Fig. 109).

The foundations of the dam consisted of nearly horizontal strata of loamy limestone with frequent layers of peat and clay. The strata were assigned on angle of friction $\phi=25°$.

The rock strata were reproduced with a mixture of pumice and cement and between the strata sheets of asbestos cardboard were interposed, these being saturated with mineral grease and frequently interrupted to impede continuity.

The primary purpose of this investigation was to check the stability against sliding of these strata for hydrostatic loading, as well as to examine comparatively the efficiency of various methods of attempting to improve the factor of safety against sliding.

Fig. 110 illustrates the principal stages of the research carried out on two basic models.

On the first model the efficiency of two solutions, that proposed anchoring the dam by means of prestressed cables leading to an inspection passage deep in the rock, were examined. The compressibility and lack of strength of the rock demonstrated the inadvisability of excessive concentrations of stress in relatively limited zones.

On the same model a concrete wedge shaped "key" was reproduced penetrating deep into the strata, to be achieved in practice by the excavation of a tunnel along the central axis of the contour of the abutment. The experiment showed that in such a position the wedge, rather than forming an anchoring structure, behaved as a rigid nucleus around which the whole structure tended to rotate as if about a pivot.

On the second model the wedge was moved downstream to receive and better distribute the flow of the compressive stresses that the dam transmits to the foundation.

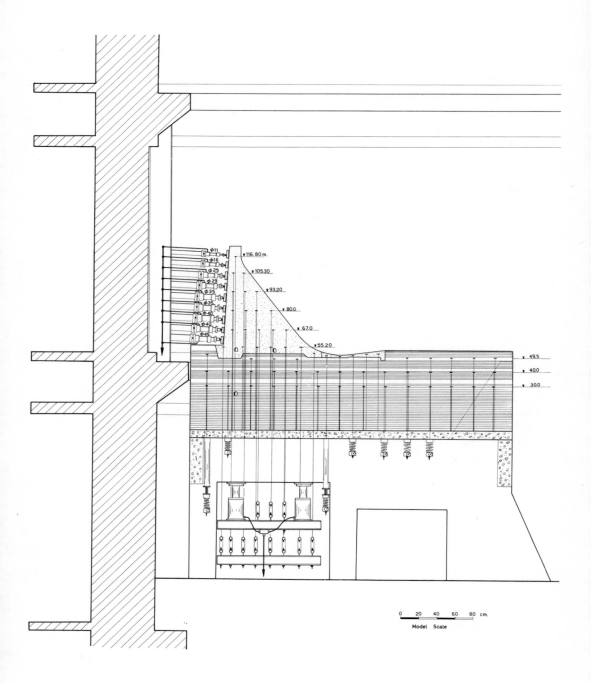

Fig. 109. Two-dimensional model of "Mequinenza" dam (Spain). Diagram of the model in the testing tower showing schematically the system for applying self weight and hydrostatic loads

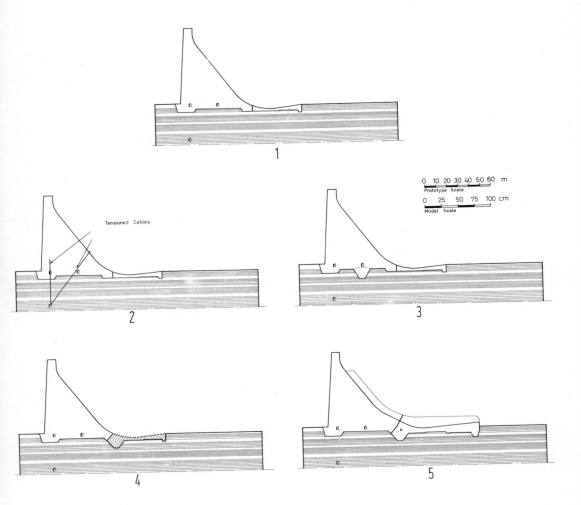

Fig. 110. Two-dimensional model of "Mequinenza" dam: (1) original model, (2) model with vertical and inclined prestressing cables, (3) model with central foundation wedge, (4) model with foundation wedge at toe of dam with replaced toe zone, (5) model with foundation wedge at toe and increased spillway section

During the experiments the behaviour of the structure was checked when subjected to partial impoundment during the excavation procedure.

This solution was shown to be effective, practical and safe.

Finally, to improve the hydraulic profile of the spillway, the concrete slab at the toe of the dam was reinforced and extended. The tests to failure showed that, the final solution increased the margin of safety of the structure against sliding by more than 50% (Figs. 111 and 112).

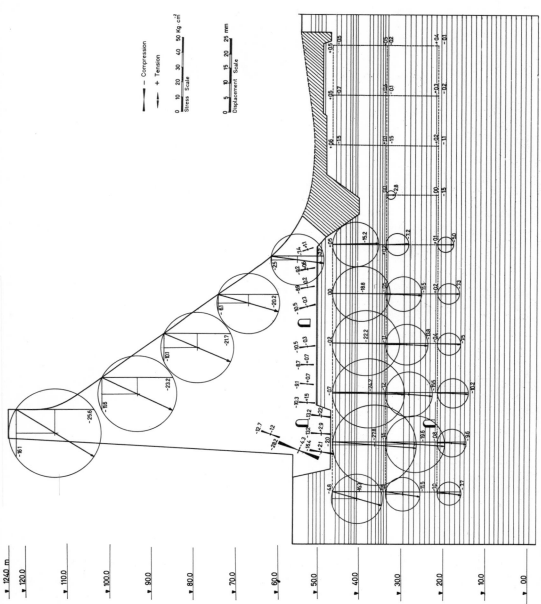

Fig. 111. Two-dimensional model of "Mequinenza" dam. Deformations and principal stresses due to self weight

Fig. 112. General view of two models of the "Mequinenza" dam (Spain)

6.3.5 Determination of the Factor of Safety

The thing that the model can provide in addition, when compared with theoretical procedures, is a check on the static behaviour of the structure up to failure beyond the limits of elasticity and also a means of ascertaining the real overall factor of safety of the dam.

This argument requires a specific explanation.

The analyses performed on the result of the experiments on a model of the "Susqueda" dam (Spain) provide a convenient practical example [12]. The model was reproduced at scales $\lambda = 120$ and $\xi = 2.7$. At the outset it is important to realise that the factor of safety varies considerably in relation to the way in which it is defined.

There is, in the first place, the traditional concept of safety as the comparison between the maximum local value of the stresses obtained in the elastic range and the failure load obtained from a uniaxial compression test on a sample.

In our case, this criterion is examined in relation to the predictable compressive strength of the concrete. Taking a uniaxial compressive strength of 390 kg/cm^2 and calculating by superposition the maximum compression as being approximately 55 kg/cm^2 then the resulting ratio of safety is equal to $n = 7$. Nevertheless, only through experiments on the entire structure does one completely exceed these ultimate loads in the evaluation of the maximum safe load, subjecting the model to gradually increasing loads up to the onset of collapse. In the present case, during these ultimate tests, the self weight load was maintained at a normal value[10], whilst increasing only the hydrostatic load.

---

10 Dealing with a relatively thin shell, testing by incrementing proportionally the two loads was not considered to be of particular interest.

The structural model is able to take into account in some measure, at least as far as the material reproducing the concrete of the dam is concerned, departures from proportionallity following the development of plastic phenomena in the dam and the rock abutments and to take account of the possibility of internal reinforcement of the structure by the high intrinsic indeterminacy.

This prevents the possibility of a direct comparison with the traditional methods of calculation but with this test to failure one gains a wealth of information in extension of the scheme as designed.

The value of the factor of safety thus obtained is certainly closer to reality than that found from the above mentioned local criterion and is normally, providing that no foundation settlement occurs, also more favourable since the ability of the concrete to adapt plastically produces a redistribution of stresses before they reach failure level.

Besides, even when this has occurred locally, it is not in general important for the collapse of the structure (which is on the contrary assumed in the criterion of local evaluation).

Therefore we can deduce that normally a test to failure on a model, in which the ratio of efficiency between model and prototype materials is held constant up to failure, provides a more realistic value of the factor of safety than that obtained by traditional calculation.

Having in particular stuck to the normal values of the self weight loads the balance between the two basic load components is substantially modified in an unfavourable sense.

Indeed, the bands of tensile stress on the upstream and downstream faces found for the hydrostatic load become particularly apparent with the reduction of compensating stresses due to self weight.

During the test to failure on the static model it was observed that up to a hydrostatic load equal to $n=6$ times normal, the behaviour of the model remained basically in the elastic range with limited permanent deformations on unloading.

At this load a primary fissure developed on the downstream face along the arch at level 310 m. On increasing the load further other fissures appeared at lower levels for a value $n=7.1$ accompanied by plastic crushing along the line of the abutment.

6.3.6 Considerations on the Mechanics of Failure of Models of Arch-Gravity Dams

Let us consider in general a model of a concrete arch-gravity dam and analyse the development of the static behaviour of the structure during a test to failure [13].

It is well known that for models loaded hydrostatically the deformation curve of the cantilevers is influenced by the conditions of constraint at the base and support at the crest.

The effect of the support at the crest introduces in particular an inversion of the normal curvature of a free cantilever in free deformation (Fig. 113) [5].

If the flexibility of the perimetral fixity at the abutment is increased, a translation of the line of the points of inflection towards the periphery is

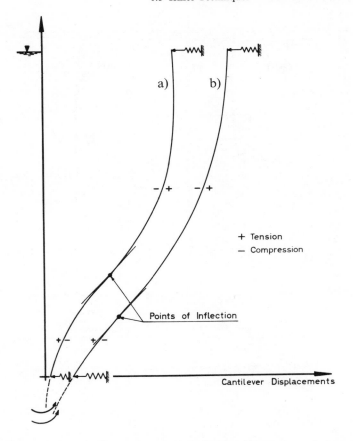

Fig. 113. Deformation curves for the crown cantilever a) for rigid perimetral constraint by the rock,   b) for flexible perimetral constraint by the rock

obtained relative to the deformed surface, together with a softening of the curvature and a reduction of the tensile stresses downstream. This latter related in effect to compressive stresses due to the self weight not entirely balancing the tensile stresses referred to above.

In this case a calculation performed by increasing proportionally the two fundamental load components (hydrostatic thrust and self weight load) permits the line of maximum tension, along which for given load conditions a separation crack will be produced, to be easily identified (Fig. 114).

Finally if the investigation to failure is effected in a conservative manner, that is to say if the self weight remains fixed at its normal value and the increase is limited to the hydraulic thrust (the only thing that in reality can increase in the event of an accident), the magnitudes of the tensile stresses increase more rapidly.

In this case it is possible to observe the fissure being produced at a lower level than in the previous case and at a reduced hydrostatic thrust.

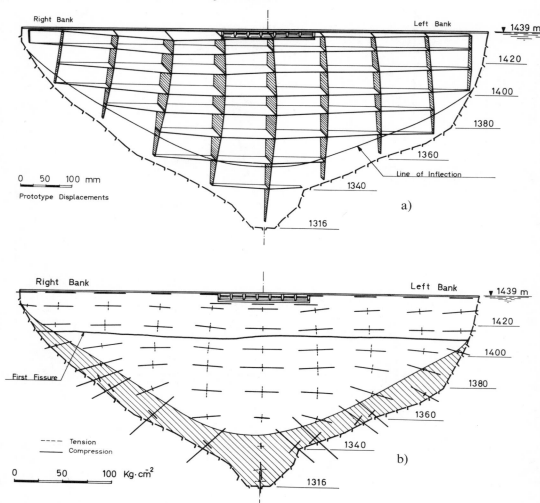

Fig. 114. Model of "Gebiden" dam (Switzerland). Hydrostatic load tests. Radial displacements and principal stresses on the downstream face: a) curves indicating the deformed surface, b) first fissure

It is observed in all cases that the fissure does not form a crack that penetrates to a definite depth, however it does not intersect the entire thickness of the dam (Fig. 115). Generally it extends rapidly in a near horizontal path along the downstream face and separates the static behaviour of the dam into two practically independent parts.

In principle the fissure does not constitute a feature prejudicial to the safety of the dam but it nevertheless substantially modifies its static behaviour. With the appearance of the fissure the lower body of the dam, by now released and independent, shows a more pronounced bending action with increased deflection downstream.

Fig. 115. Classical horizontal separation crack, "Agua de Toro" dam (Argentina)

The upper body, having lost the assistance of the cantilevers, now behaves predominantly as a horizontal arch. As a consequence there is an important increase in the forces applied to the lateral abutments.

If the rock abutments at the upper levels are sufficiently large the structure still shows a residual strength. If this is not so then the fissure is followed by a rapid and general failure of the upper body of the dam through settlement of the foundation.

In this context the static analysis of the model in the elastic regime generally shows a "dipping" arch type of behaviour and shows up the greatly reduced stress levels at the supports of the upper arches and in the artificial buttresses at

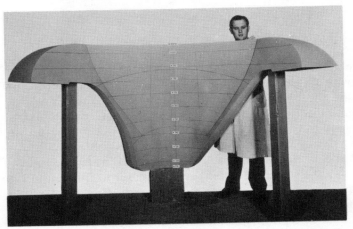

Fig. 116. Preliminary wooden model of the "Soledad" dam (Mexico)

10*

the crest (if they exist). Such evidence has led to the static contribution of the above mentioned buttresses being underestimated and to their dimensions being reduced to an appreciable extent.

However, on the basis of the failure mode examined above, the large supporting buttresses re-establish their importance, both as massive anchorages and as structures capable of providing a regular diffusion of the forces from the shell into the supporting rock abutments (Fig. 116). Frequently only the first of these functions has been considered, the second one being neglected.

Fig. 117. Failure in the buttress at the abutment. Model of "Almendra" dam (Spain)

For economic reasons the buttresses are frequently in the form of massive structures with abrupt variations of section, particularly at the junction between the shell and the buttresses. The rapid diffusion of the main stress flow on passing from the slender shell to the enlargement of the buttress induces significant tensile stresses along the surfaces of intersection, normal to the main stress field; during the tests to failure these tensions can initiate separation between the contacting structures in the vicinity of the enlargements (Fig. 117).

In this way the statically useful part is reduced to the central area defined by the surfaces that observe a smooth diffusion of the stresses from the dam to the rock abutments.

It is not claimed, a priori, that the failure mechanism examined here is the only one possible, but it is certainly the most common one.

It must be considered classical, particularly when the dam-rock abutment system is correctly assessed and as a result has the greatest reserve of strength.

If the rock along the line of the abutments is relatively weak it may follow that (taking account of the weakness at the edges) the separation crack is retarded; in this case it is possible that with the appearance of this fissure the body formed by the lower part of the shell settles rapidly due to general movement at the boundary.

In the limit, for very weak rock, settlement at the boundary can also occur before the separation crack.

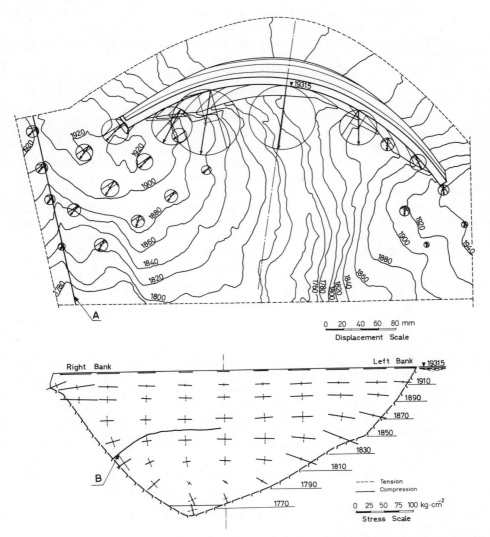

Fig. 118. Model of "Emosson" dam (Switzerland). Displacements and principal stresses on the downstream face for hydrostatic load: (A) joint open, (B) first fissure

On the other hand, when a well designed thin dam is contained in a very rigid gorge, instataneous failure can occur due to instability at exceptionally high load levels. The experimental results obtained for the "Emosson" dam (Switzerland) (Fig. 118) represent an example of particular interest [5].

It is well known that the Boussinesq theory, used to calculate the deformations at the dam supports, assumes that the rock extends indefinitely from the dam abutments.

In the case under examination the rock buttress on the right bank is practically isolated by a rather important open fault and is defined on the downstream side by contours that fall rapidly to a relatively low level.

The rock mass is sound with some joints and was reproduced in the model so that it had the same average modulus of elasticity as the concrete.

It behaved in effect as a relatively slender and elastically rather deformable buttress.

As a result the deformations measured at the abutments were sensibly assymetric, being greater at the right bank; thus differing from those predicted by the theoretical analysis.

The deformability of the abutments introduced torsional forces in the lateral blocks on the right bank with considerable development of tensile stresses parallel to the line of the abutment along the downstream face.

A primary fissure was found intersecting the whole thickness at the location of these tensile stresses at a much lower value of the hydrostatic load than should have been expected.

To improve the factor of safety and to some extent the symmetry of the behaviour of the dam it was necessary to design a somewhat substantial increase of the upstream thickness of the dam for the whole region on the right-hand side.

The static behaviour of the new design was shown to be completely satisfactory by subsequent tests on a model thus modified.

References

[1] Model Test of Boulder Dam, Part V, Technical Investigations, Bulletin 3. Bureau of Reclamation, Denver, Colorado (1939).
[2] Azevedo, M. C., Ferreira, J. E.: Construction of Models of Concrete Dams for Elastic Tests. Memoria L.N.E.C., Nr. 232, Lisbon (1964).
[3] Beaujoint, N.: Modèle réduit en cautchouc du barrage de Roselend. Colloq. Intern. sur Modèles Réduits de Structures, Madrid (June 1959).
[4] Fumagalli, E.: Stability of Arch Dam Rock Abutments. Bulletin Ismes Nr. 32 (May 1967).
[5] Fumagalli, E.: Influence des fondations sur la mécanique de rupture des barrages-voûte. Proceedings 2nd Congress of I.S.R.M., Beograd, 1970, Bulletin Ismes Nr. 46 (October 1970).
[6] Lauletta, E.: Thermoelastic Tests on Arch Dam Models. Bulletin Ismes Nr. 27 (October 1964).
[7] Nizery, A., Remenieras, G., Beaujoint, N.: Etude sur modèle réduit des contraintes dans les barrages. Annales des Ponts et Chaussées (July-October 1953).

[8] Oberti, G.: Essais sur modèles des barrages. Wasser- und Energiewirtschaft Nr. 7, 8, 9, 1956. Bulletin Ismes Nr. 7 (July 1957).

[9] Oberti, G.: Arch Dams: Development of Model Researches in Italy. Bulletin Ismes Nr. 9 (December 1957).

[10] Oberti, G.: Italian Arch Dam Design and Model Confirmation. Bulletin Ismes Nr. 14 (March 1960).

[11] Oberti, G.: Modelos de presas de concreto y tuneles. Bulletin Ismes Nr. 37 (December 1967).

[12] Oberti, G., Fumagalli, E.: Sul funzionamento statico della diga di Susqueda dall'analisi dei risultati sperimentali su modello. L'Energia Elettrica Nr. 7, Vol. XLVII (1970). Bulletin Ismes Nr. 47 (November 1970).

[13] Oberti, G., Lauletta, E.: Evaluation Criteria for Factors of Safety. Model Test Results. Bulletin Ismes Nr. 25 (October 1964).

[14] Réméniéras, G.: Quelques études et réalisations Françaises récentes dans le domaine des modèles structuraux. Convegno di Venezia su „I modelli nella tecnica", Vol. 1 (October 1955).

[15] Roberts, A.: A Model Study of Rock Foundation Problems underneath a Concrete Gravity Dam. British Society for Strain Measurement, Vol. 1, Nr. 3, 4—9 (July 1965).

[16]. Rocha, M.: Structural Model Techniques: Some Recent Developments. Memoria L.N.E.C., Nr. 264, Lisboa (1965).

[17] Rocha, M., Serafim, J. L.: Rupture Studies on Arch Dams by Means of Models. Water Power (1959).

[18] Rocha, M., Serafim, J. L., Fernandez, A. G., Poole Da Costa, J.: Experimental Studies of Buttress and Multiple Arch Dams. Septième Congrès des Grands Barrages, Rome (1961).

[19] Rocha, M., Serafim, J. L., Ferreira, M. J. E.: The Determination of the Safety Factor of Arch Dam by Means of Models. Proceedings, Symposium on Models of Structures, Madrid, 1959, Rilem Bulletin, Paris, New Series Nr. 7, 68—78 (June 1960).

[20] Rocha, M., Serafim, J. L., Silveira, A. F.: Deformability of Foundation Rocks. 5th Congress on Large Dams, Report Nr. 75, Paris (1955).

[21] Rocha, M., Silveira, A. F.: The Use of Models to Determine Temperature Stresses in Concrete Arch Dams. Paper Nr. 23, Symposium on Concrete Dams Models, L.N.E.C., Lisboa (1963).

[22] Rocha, M., Silveira, A. F., Cruz Azevedo, M. C.: Note on Some Comparison Between Experimental and Analytical Values of the Stresses and Displacements of Concrete Dams. Symposium on Concrete Dams Models, Lisboa (October 1963).

[23] Serafim, J. L., Da Costa, G. P.: Methods and Materials for the Study of the Weight Stresses in Dams by Means of Models. Bulletin Rilem, Nr. 10 (March 1961).

[24] Sparkes, S. R., Chapman, J. C., Cassel, A. C., Chitty, L., Hobbs, R. E.: Model Analyses of an Arch Dam with Different Valley Stiffnesses. Imperial College, Technical Report AD/5 (September 1965).

7. Geomechanical Models

7.1 Geomechanical Models of Dams

A greater understanding of the mechanics of large rock systems has been shown to be of great moment in recent years in connection with the stability of foundations and underground excavations for major civil and industrial projects. However, without belittling interest in problems associated with the construction of large bridges, underground power stations, pressure tunnels and transport tunnels, it is worth noting that the study of and research into the stability of artificial hydraulic basins has time and again acquired an even greater relevance because of the importance that attaches to the safety of such structures.

The problem consists of investigating with a model the whole system of the dam and its confining rock mass when subject to the combined action of the self weight of the entire system and the thrust produced by the impounded water against the dam and against the rock mass.

Experiments of this kind must be developed on predominantly visco-plastic models through gradually increasing load cycles, which by their nature are non-repeatable, up to final collapse of the system.

Working according to equation (10) at the ratios $\zeta \simeq \lambda$, the self weight is normally reproduced to a scale $\rho \simeq 1$ by the density of the materials that produce the gravitational stress fields in such cases.

Considering that the investigations are mainly influenced by the behaviour of the confining rock mass, the question arises as to whether it is possible, and to what extent, to reproduce the complexities of the rock mass under experimental conditions.

In this context one aspect deserves to be clarified; it concerns the current position and significance that these geomechanical models acquire in the overall design process of the project. Having completed the geological and geomechanical investigations on the rock mass, the geologist prepares a geomechanical report; on the basis of this report the designer, or in some cases the same geologist, prepares a structural idealisation of the rock mass necessary for investigating its equilibrium.

Mathematical methods have in general been shown to be inadequate for the analysis of such complex three-dimensional schemes. An investigation on a geomechanical model is generally more suitable, the validity of this being strictly related to the correspondence between the idealised scheme and the real situation.

Logically, because of the inevitable shortage of reliable data, the model must be designed with extreme caution, so that in no case does it reproduce conditions more favourable than those of the prototype.

In the orbit of idealisation of this type it is evident that the geomechanical model produces more limited margins of safety than conventional static models.

Nevertheless, because of the extreme caution exercised, tending to guarantee every aspect of the project and its foundation from eventual disaster, the value of the factor of safety so obtained assumes the significance of a lower limit below which it is not humanly forseeable to descend.

7.1.1 Techniques of Modelling and Experimenting

Because of their importance in the construction of models, it is necessary to observe that any rock mass contains systems of discontinuities of varied nature and importance (schistosity, joints, stratification, faults, cavities, etc.). Whereas the more important discontinuities, such as major faults, should for the most part be individually reproduced, or at least with reduced frequency as in the case of masses which have known and suitably spaced planes of discontinuity (e.g. joints or bedding planes); the minor discontinuities are normally lumped in with the general properties of the rock mass, particularly when their distribution does not induce anisotropy in the rock.

To take account of such distributed properties in the model an adequate reduction is made in the mechanical properties of the rock matrix.

Details of suitable types of material for reproducing stone materials at scale values ζ between 100 and 150 were given in section 2.3.4.

It is now necessary to examine the techniques required to reproduce the planes of discontinuity (schistosity, joints). The most widely used method is that of moulding the material into small bricks. For this purpose a hydraulically operated briquette press, together with a limited air drying of the bricks before placing in the model, produces very satisfactory results [4].

Besides being able to place the ready made briquettes according to predetermined facial planes, one can produce in the mass three and sometimes four separate joints sets.

It is, however, important to realise that in positions were close contact is required, the briquettes generally produce excessively large gaps between adjoining surfaces when compared with the prototype.

It follows that in order to avoid excessive reduction in the overall modulus of deformability of the system, the dimensions of the briquettes must never be too small and the resulting frequency of discontinuities is then notably less than that found in the prototype. In each case the dimensions of the briquettes and therefore the interval between the joints are checked by testing the overall behaviour of an assemblage of at least one hundred briquettes (Fig. 119) [2].

Fig. 119. Overall determination of rock deformability

Such an experiment allows one to construct a diagram of the type shown in Fig. 120 relating to the model rock for "Rapel" dam, showing satisfactory correlation with Fig. 3.

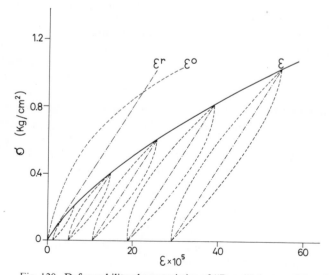

Fig. 120. Deformability characteristics of "Rapel" dam model rock

Fig. 121. Simulation of "Emosson" rock by blocks cast in-situ

When the in-situ rock is compact and has a high modulus of deformability the joints are reproduced directly in the model by casting the surfaces in close contact (Fig. 121). Continuity in this case can be avoided by coating the surface with isolating varnish.

Consider now the frictional and deformability conditions relative to an important discontinuity, such as a fault or bedding plane (ignoring for the time being pore pressure). The situation is substantially affected by the presence or otherwise of infillings of clay, peat or mylonites. In the absence of infillings the angle of friction is little different from the angle of internal friction and, for reasons of prudence, a value between $30° \leq \phi \leq 45°$ is normally adopted in the model.

The irregularities of the rock surfaces are normally neglected in the model. In fact for simplicity they are for the most part idealised and cautiously reproduced as plane surfaces. In the presence of infillings of clay or peat it is necessary to consider whether the infilling constitutes a true layer, forming a continuous plane of sliding without contact between the asperities or irregularities of the opposed rock surfaces.

In this case the angle of friction is closely related to the moisture content of the material. It is known in fact that for saturated clays the cohesion is greatly reduced and the angle of friction may be reduced to a value $\phi < 10°$. Table 7 gives details of materials and treatments suitable for reproducing different angles of friction. In general the materials that are inserted dry, such as talc, dust,

Table 7. *Infilling Materials between Sliding Surfaces to Produce Different Angles of Friction*

Infilling material	Angle of friction
Sheets of tin foil smeared with molibdenum di-sulphide	5°— 6°
Lining of alcohol based varnish and grease	7°— 9°
Lining of alcohol based varnish with grease and talc in varying proportions	9°—23°
Lining of alcohol based varnish with talc	24°—26°
Polythene sheet in 2 or 3 layers	24°—26°
Lining with alcohol based varnish	32°—37°
Powdered limestone	35°—37°
Untreated surfaces in contact	38°—40°
Sand in different grain sizes	40°—46°

graphite, etc., are better suited to reproducing the plastic sliding typical of mylonites whilst greasy paraffin based substances are better for reproducing viscous sliding conditions at low angles of friction typical of layers with clay infillings.

In those cases where the thickness S of the introduced material of modulus E is not negligible the effect on the overall deformability in the direction normal to the plane of stratification, can be satisfied by reproducing the quantity $Z = E/S$ kg/cm^3 in the model with a value, according to equation (10), of:

$$Z' = Z\frac{\lambda}{\zeta}.$$

Frequently the joints are not continuously open and the percentage ratio between closed and open surfaces is evaluated statistically by the geologist. In this case the surfaces of the briquettes in the model are also joined with glue in the same ratio.

If the glue line happens to be somewhat stronger than the material, an eventual failure can equally occur during the test as soon as continuity of the joint plane is achieved along the side of the glue line. In all cases it is advisable to use the thinnest film of glue possible.

In this context it is worth noting the shear stress-sliding displacement curves shown in Fig. 122. Diagram A relates to sliding between unbonded contacting surfaces. The initial peak represents the initial cohesive contribution of the bonded surfaces [2].

Finally if a sliding plane contains areas of varying roughness, the effective angle of friction for the whole surface does not correspond to the weighted mean value, but to a sensibly higher value. This phenomenon, found experimentally, is due to the stress concentrations that arise in the rougher zones in the phase preceding sliding.

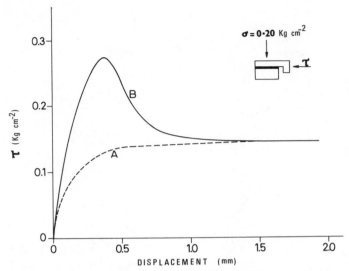

Fig. 122. Sliding versus shearing stress diagrams. Curve A, unglued specimen; curve B, specimen glued for one-third of its surface

The hydrostatic load on the model is normally applied by means of bags of liquid, using the liquids already discussed in section 3.4.1, up to densities equal to $\gamma = 3 \ t/m^3$.

Mercury clearly has too high a density to be used easily in association with the others already described.

Greater densities are normally obtained with suitable systems of hydraulic jacks and distribution plates, much lighter than those used for conventional models.

In geomechanical models of dams over loads relative to flood waves and — to a certain extent — effects resulting from explosions, landslips and seismic disturbances can be reproduced by increasing the level of the bags beyond the normal depth.

Again in analytical procedures one in fact adopts, to a first approximation, overload envelopes of uniform intensity. In this case, to investigate the behaviour of the mass subject to the whole load of the basin, a large part of the wetted orographic surface is covered with one or more bags of liquid.

On the assumption of relevant fissuration effects the hydraulic thrust is transferred directly to the impermeable curtain. The hydraulic thrust can be applied as the total pressure due to the compounded water or reduced on the more reasonable assumption of pressure losses during percolation of the water. In this case the rock upstream of the impermeable curtain is entirely removed.

Sometimes it is of interest to investigate the stability along planes of possible detachment or sliding due to hydrostatic uplift. For this purpose pneumatic bags filled with sand as shown in Fig. 123 have given good results. The sand in the bags has the principal task of guaranteeing continuity of contact in the material without impeding the free diffusion of the compressed air.

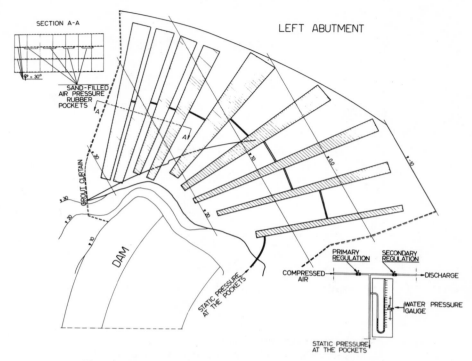

Fig. 123. "Rapel" dam model. Layout of uplift simulating bags

The system for regulating the air pressure is also shown in Fig. 123.

Finally by adopting lower angles of friction for the surfaces with bags than for those without bags it is possible to achieve, on increasing the air pressure, gradually reducing frictional conditions at the same time as load is transferred from the surfaces without bags to those with pressure bags.

To a certain extent this reproduces the reduction of the angle of friction of the infilling material between the strata that follows on the infiltration and percolation of water.

The more important measurements to be made on the model are those of absolute and relative displacements. In particular, since the anelastic deformations develop with time at constant load, the use of recording apparatus is extremely advisable.

Some limited measurements of the stress levels can be obtained from the model of the structure and in every case the significance of these values is purely indicative, considering that highly deformable materials are not very suitable for the application of extensometers.

These measurements become more difficult in a discontinuous rock system in which the stress field varies in continuity because of the irregular development of anelastic subsidence of the abutments.

Measurements of absolute displacement are for preference taken on surfaces as, where possible, three components of spatial displacement.

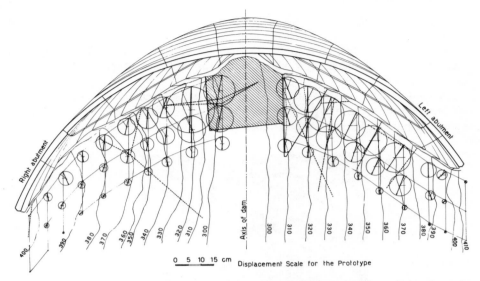

Fig. 124. "Grancarevo" dam geomechanical model. Rock surface displacement circles along the downstream abutments during the failure test

On relatively flat areas of the rock surface one can set up a triangular and rectangular grid of demountable deformation gauge points. Taking measurements with respect to a stable reference point, so as to measure absolute displacements, the displacement pattern over the whole grid can be established (Fig. 124). Making measurements within the grid gives the local relative deformations which can then be correlated with the absolute displacement values.

Within the rock mass, across the more important discontinuities, special deformation gauges placed in a St. Andrews cross configuration permit the

Fig. 125. Deformation meters for measurement of internal relative displacements

determination of the relative movements of the sliding surfaces and the separation of these surfaces. These deformation gauges consist of electrical resistance strain gauges attached to a highly deformable rubber support. The support is fixed to two metallic anchor plates which anchor the gauge to the material (Fig. 125).

Measurements of absolute components of displacement can still be made within the rock mass using reference "invar" wires suitably anchored within the rock and passing freely to the surface through small boreholes.

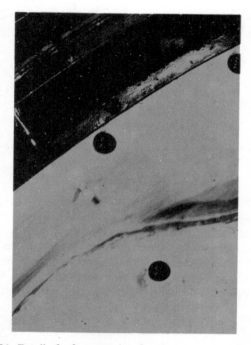

Fig. 126. Detail of reference points for photogrammetric survey method

Of considerable interest, particularly in problems involving large deformations and tests to failure, are photogrammetric surveying techniques. Photographs before and after loading are taken on the same photographic plate with the camera fixed rigidly at a distance of some metres from the model. The displacements of sufficiently clear and distinct reference points can be evaluated to an accuracy of ±0.1 mm (Fig. 126). A survey conducted with two cameras operating in orthogonal planes can provide, after suitable analysis, the spatial displacements of the reference points.

7.1.2 Some Examples of Models and Analyses of Results

Table 8 lists the geomechanical models of arch dams that have been built and tested at Ismes using the methods described here.

Table 8. *Results of Testing Geomechanical Models*

Model	λ	ζ	ρ	Construction of rock mass	Discontinuites	Young's modulus of prototype (kg/cm²)		Hydrostatic load
						Homogeneous rock	Overall of rockmass	
Vajont (Italy)	85	85	1	3200 precast oblique prismatic bricks, 16 × 14 × 9 cm	Faults and bedding planes	600,000	200,000	With liquids
Pertusillo (Italy)	150	100	0.66	Blocks cast on iron	Fractured bedding planes	350,000	25,000—50,000 (anisotropic rock). 10,000 (grès)	With liquids and jacks
Kurobe (Japan)	100	50	0.5	Continuous casting in-situ	13 faults and weak zones	—	12,700—72,500	With liquids
Ca'Selva (Italy)	80	128	1.6	8000 precast oblique prismatic bricks of two types for left bank, 10 × 10 × 10 cm	Bedding planes and 15 faults	320,000	40,000+80,000	With liquids and jacks
Grancarevo (Yugoslavia)	120	80	0.66	12000 precast oblique prismatic bricks, 5 × 10 cm, $h = 3$—6 cm	Bedding planes and 14 faults	480,000	Anisotropic rock 70,000—250,000	With liquids and jacks
Emosson (Switzerland)	100	150	1.5	Oblique blocks cast in-situ 10 × 10 × 40 cm	2 faults	400,000	300,000	With liquids
Rapel (Chile)	100	100	1	20000 precast oblique prismatic bricks of 3 types, 8 × 8 × 8 cm	7 faults	10,000 for the outer zones	70,000—160,000	With liquids
Susqueda (Spain)	120	120	1	2000 precast oblique prismatic bricks and continuous casting in-situ	10 faults	—	75,000—150,000	With liquids and jacks

It must be recorded that the experiments on the geomechanical models of "Vajont" dam suggested the appropriate strengthening measures to assure the stability of the overhanging walls of the abutments, measures subsequently shown to be correct and efficient to ensure the safety of the dam during the tragic slip failure of mount Toc.

The tests on the model of the "Pertusillo" dam produced extremely interesting research results, particularly in connection with the stability and stiffness of its foundations; correlating with the hypotheses and theoretical analyses developed in this context by the author [5].

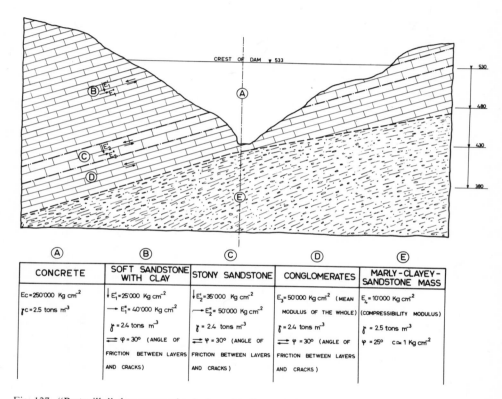

(A)	(B)	(C)	(D)	(E)
CONCRETE	SOFT SANDSTONE WITH CLAY	STONY SANDSTONE	CONGLOMERATES	MARLY - CLAYEY - SANDSTONE MASS
$Ec = 250'000$ Kg cm^{-2}	$\downarrow E_1' = 25'000$ Kg cm^{-2}	$\downarrow E_2' = 35'000$ Kg cm^{-2}	$E_3 = 50'000$ Kg cm^{-2} (MEAN	$E_4 = 10'000$ Kg cm^{-2}
$\gamma c = 2.5$ tons m^{-3}	$\longrightarrow E_1'' = 40'000$ Kg cm^{-2}	$\longrightarrow E_2'' = 50'000$ Kg cm^{-2}	MODULUS OF THE WHOLE)	(COMPRESSIBILITY MODULUS)
	$\gamma = 2.4$ tons m^{-3}	$\gamma = 2.4$ tons m^{-3}	$\gamma = 2.4$ tons m^{-3}	$\gamma = 2.5$ tons m^{-3}
	$\rightleftharpoons \varphi = 30°$ (ANGLE OF	$\rightleftharpoons \varphi = 30°$ (ANGLE OF	$\rightleftharpoons \varphi = 30°$ (ANGLE OF	$\varphi = 25°$ c $\approx$ 1 Kg cm^{-2}
	FRICTION BETWEEN LAYERS	FRICTION BETWEEN LAYERS	FRICTION BETWEEN LAYERS	
	AND CRACKS)	AND CRACKS)	AND CRACKS)	

Fig. 127. "Pertusillo" dam geomechanical model. Cross-sectional geological sketch as simulated in the model

Fig. 127 shows a transverse section through the rock model indicating the stratification of the marley-clayey-sandstone foundation that exhibited the mechanical properties of a highly compressed incoherent soil (more than of a rock).

Similarly the experiments on the geomechanical model of the "Kurobe" dam have been shown to be of great usefulness, particularly with reference to the investigation of the equilibrium of the rock abutments in the zone below the crest and to the more appropriate precautions for reducing the thrust of the

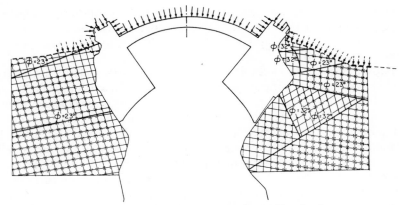

Fig. 128. "Rapel" dam model cross-section at 60 m level

upper arches transmitted, through the spatial arching action of the structure, to the lower levels.

Technically the most advanced model, both for techniques of construction and manner of testing, is that of "Rapel" dam, modelled on information provided by Muller (Fig. 128). In this a pressure load application system using pneumatic bags, already described in Fig. 123, was used for the first time. The system was adopted for five different individual sections, that were among the more dangerous, in the buttress at the left bank.

Turning now to examine the processes of deformation and failure, it is necessary to realise how complex the intrinsic structure of the rock mass can be and to differentiate clearly between the behaviour of surfaces subject to uniaxial and biaxial stresses and those subject to triaxial stresses. In the former case the strength is principally due to cohesion leading to a basically brittle type of failure, but at depth, in the presence of isotropic stress states, the strength is principally related to the angle of internal friction and failure, of the plastic type, occurs only in the presence of high shearing forces.

It is also observed that whereas the near surface zones of a rock mass behave predominantly in an elastic manner, at depth plastic behaviour predominates with the occurrence of large deformations, during which the system maintains, through friction, appreciable strength reserves despite rupture of the cohesive bonds.

In particular it should be noted that at the surface the cohesive bonds can make little use of the contribution due to mutual collaboration between the many possible load paths in the system and fracture occurs abruptly as soon as the elastic limit is passed.

At depth on the other hand, as the load increases, the basically plastic deformation processes affect the resisting structure in successive boundaries at ever increasing depths and distances from the surface of application of the forces.

With the retreat of this boundary the cohesive bonds gradually yield while the resistant tensions, associated with the internal friction of the material, are called on to mutually assist one another in a gradual stress redistribution, all of which occurs logically in the regime of large deformations.

Under constant load, as long as a possible static equilibrium configuration exists, the velocity of the deformation processes decreases and in time becomes zero; the more slowly the closer the structure comes to the conditions of final collapse.

Finally the question arises as to what are the conditions of instability that obtain at the discontinuities present in the rock mass (joints, faults, failure planes, stratifications, etc.). If the examination is limited to considerations of a general nature it will be noted that, through lack of cohesion and of an effective isotropic triaxial stress near the surface, the dangers of instability at the discontinuities are much greater the nearer these are to the surface. When, on the contrary, the discontinuities penetrate deeply into the rock, stability depends mainly on the angle of friction between the contacting surfaces.

When the value of this angle is approximately equal to that of the internal friction of the rock (rock actively in contact), sliding of these surfaces, no matter how important and extensive they are, occurs only in exceptional cases; for which the absence of rigid cohesive bonds (destined, despite what has been said, to be gradually broken by successive boundaries) does not normally constitute an element of known weakness in the system.

Fig. 129. "Ca'Selva" dam model after failure tests

Only in the case in which the angle of friction between the surfaces of the discontinuities is less than the angle of internal friction does failure occur preferentially by sliding of strata as in the case of the geomechanical model of "Ca' Selva" dam (Fig. 129).

The worst conditions normally occur where there are infillings of wet clay or peat. In such a case sliding can take place at angles of friction $\phi < 15°$. A case

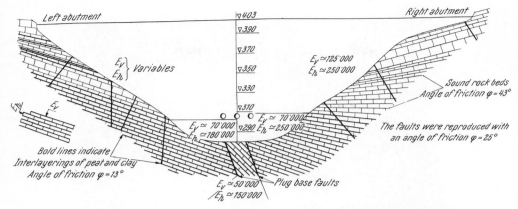

Fig. 130. Schematic geomechanical sketch of the "Grancarevo" dam foundation rock

of this type occurred for the "Grancarevo" dam in Bosnia Herzegovina. The abutments consisted of layers of calcareous rock sometimes interlayered with clay and peat. The model was reproduced according to the scheme shown in Fig. 130 with prismatic briquettes of variable thickness (Fig. 131), working on suitably matched beds, strata by strata [2].

The test to failure on the model showed the sliding on each of the infilled layers (Fig. 132).

A rigorous investigation on the model allowed an efficient system of strengthening the strata, using stressed cables, to be adequately designed.

Fig. 131. "Grancarevo" dam. Rock modelling by bricks

Fig. 132. "Grancarevo" dam model after failure tests

In Fig. 133 can be seen the pattern of displacements obtained from the geo-mechanical model of "Vajont" dam for normal impoundment.

Analysing these displacements into components of displacement intrinsic to the arches and components of free translation as shown in Fig. 134, it can be seen that only the former induce forces in the arches, whilst the latter can be considered at best as rigid body movements.

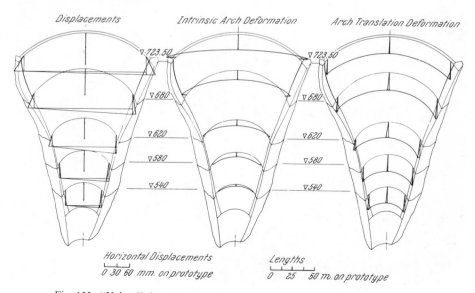

Fig. 133. "Vajont" dam model displacements measured at full impoundment

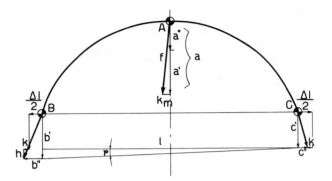

Fig. 134. "Vajont" dam model resolution of displacement pattern

Intrinsic arch deformations

$\Delta 1 = b'' - c''$

= elongation of chord 1

a'' = radial deflection at crown of arch

$k_m - k$ = lateral intrinsic displacement measured at the crown

Translation displacements

b', c' = translation components of points „B" and „C" perpendicular to chord 1

$a' = \dfrac{b' + c'}{2}$ = radial translation component at the crown

$k = \dfrac{b'' + c''}{2}$ = lateral translation component along chord 1

$\varphi = \dfrac{h}{1}$ = rotation angle referred to point „C" due to translation movements

$h = b' - c'$

The major part of the rigid body movement consists of a vertical displacement downstream of the order of 40 mm in the direction of the axis of the valley.

More limited, and proportional to the rigidity of the arches at the individual levels, are the displacements of intrinsic deformation.

Fig. 135 shows a comparison between the components of intrinsic radial deformation measured at the main section, obtained respectively from geomechanical models and from the conventional type models.

Whilst the agreement is satisfactory for the models of "Kurobe" and "Pertusillo" dams, in the models of "Vajont" dam, while noting a certain similarity in the curves of deformation, the geomechanical model shows displacements approximately double those of the traditional model.

This latter is accounted for by the greater rigidity adopted for the rock in the conventional model.

As a general observation it can be seen that the collapse conditions of the dam are more closely connected with the settlement of the abutments than with the intensity of the applied load.

In essence, as the thrust on the rock abutments increases, they dissipate the load through large deformation visco-plastic processes that gradually extend to the entire perimetral development of the abutment, whilst the dam wall — more rigid — exceeds its intrinsic capacity for deformation, in conformity with the deformations imposed by the settlement of the foundations, resulting in failure.

In confirmation of this it is interesting to observe that the maximum deformation measured at the onset of final collapse in the crown of the dam wall is the same for both geomechanical models and conventional models.

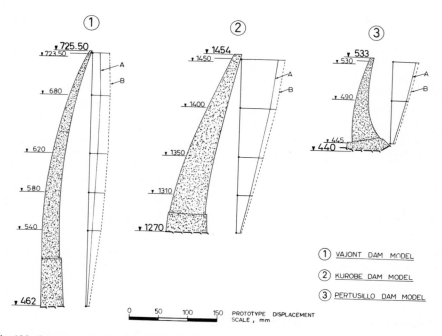

Fig. 135. Comparison of intrinsic displacements measured on conventional (A) and geomechanical (B) models. (1) "Vajont" dam model, (2) "Kurobe" dam model, (3) "Pertusillo" dam model

It is of particular interest to plot separately, as functions of the intensity of the applied load, the values of both the reversible and the residual anelastic components of the total deformations at points of particular interest. Fig. 136 shows such a plot for the radial deformations at the crest of the central cantilever for the geomechanical model of "Susqueda" dam (Spain).

At normal load ($n=1$) the deformations are mainly of a reversible nature and among other things are not much different from those measured on a conventional model. For greater loads the anelastic residues increase with the load in a strongly exponential manner, while the reversible deformations do not depart greatly from linearity.

From this it appears evident that it can be wrong and invidious to draw conclusions on the safety of the structure from an examination of the deformations

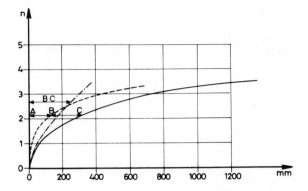

Fig. 136. Geomechanical model of "Susqueda" dam. Curve of the deformations at the crown of the crest arch. AC) Total deformation, BC) Reversible deformation, AB) Permanent deformation

at normal operating loads, whether they are obtained from in-situ measurements or on models or else from theoretical calculations, when failure occurs by settlement of the foundation.

Only from a test taken up to collapse can one obtain the information relative to the settlement of the rock mass; the one thing that allows one, through an objective examination of the real behaviour of the whole system of the dam with its rock foundation, to establish the real factor of safety of the structure.

As an illustration for "Susqueda" dam Fig. 137 shows a comparison of the radial displacements measured at the position of the central cantilever in the conventional model and in the geomechanical model for full water level in conditions of normal operation. For the geomechanical model displacements are also shown for cases of partial filling of the basin, of great use for monitoring those of the prototype during the first cycle of loading.

As a final consideration it is found that, based on many observations on the behaviour and scope of geomechanical models, they are not well suited to the determination of the stress distribution in the dam in the elastic range. They are intended mainly to investigate the static equilibrium of the whole system of the dam and the rock mass, beyond the limit of elastic behaviour in the visco-plastic regime, up to failure.

For this reason they do not replace models of the conventional type but complement them.

7.2 Two-Dimensional Geomechanical Models

In addition to the research activity on three-dimensional geomechanical models, experimental studies on plane models have been performed at Ismes and in other laboratories in the field of underground excavations, particularly for controlling the stability of tunnels, dimensioning their linings, or establishing — in the case of pressure conduits — the collaboration of the rock in the hoop action of the conduit.

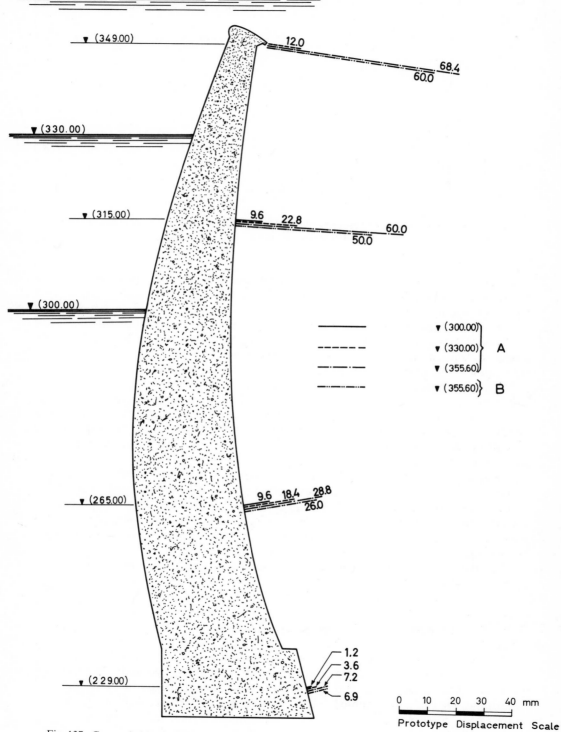

Fig. 137. Geomechanical model of "Susqueda" dam. Crown section. Comparison between radial displacements obtained from the geomechanical model (A) and from the conventional model (B)

A close collaboration between the geologist and the experimentor is again in this case of great value in choosing the geomechanical idealisation that best reproduces the in-situ conditions, taking good account of the limitations of ordinary practice.

The experiments can be developed alternatively along two lines; the one consists of producing a complete model of the tunnel and, where appropriate, its lining, before applying the loading pattern to the boundary of the rock model; the other of applying these load conditions before proceeding to reproduce the tunnel and its lining in the model.

If the organisation of the work is likely to lead to a rapid excavation procedure and immediate lining of the tunnel walls, it must be assumed that the main part of the visco-plastic effects will develop after completion of the work; in which case the first procedure can reasonably be adopted.

It is useful to note here that according to equation (10), considering the possibility of applying the force produced by the massif as a boundary condition of the rock model, the scale of the density can be varied at will; it then follows that the specific force scale ζ can be chosen freely, permitting the use of materials with higher mechanical properties than those required for three-dimensional models.

The reproduction of discontinuities in plane models is important only if the plane in which they lie is parallel to the axis of the tunnel.

It would be very interesting in some cases to proceed to experiments on a spatial model of the entire massif through which the tunnel passes, but this requires models of large dimensions frequently too onerous to be arranged for a tunnel of reasonable size.

Other difficulties result from the absence of information on the internal stress state of the rock, which can vary considerably in intensity and in direction along the tunnel, and is in general neither proportional to the thickness of the rock overburden nor exclusively vertical.

In default of this data the model investigation is carried out by subjecting the rock to simple stress states or stress states arranged according to different alignments.

A procedure of this type is useful for designing a universal lining that is suitable for all the possible pressure conditions that may occur along the tunnel. During the test the stresses and deformations are measured in the tunnel lining and, if possible, on the visible surface of the rock. More generally the disturbance of the isostatic flow in the rock mass, produced by the presence of the tunnel, is determined.

Tests to failure are finally carried out for the loading conditions that are considered to be the most important.

As an illustration Fig. 138, shows the model of the tunnel at the summit of the Cisa Pass autostrada in the Appenines [2]. The photograph shows the model being subjected to a horizontal thrust. Fig. 139 shows the pattern of the isostatics and of the stresses in the lining for a uniform applied pressure of 100 t/m² and Fig. 140 shows details of the failure in the model for loads inclined at 45°.

Fig. 141 shows a photograph of the model of the bottom discharge tunnel through the base of the "Salagon" rock fill dam (France) during testing. A diagram of the thrust of the rock fill on the lining, found from an analysis of the stresses

Fig. 138. Cisa Pass tunnel model under horizontal thrust

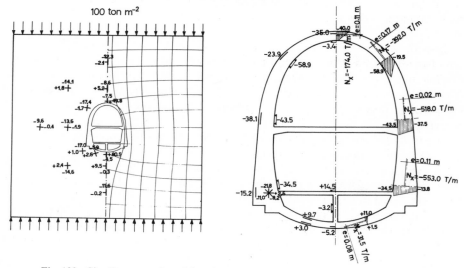

Fig. 139. Cisa Pass tunnel model under vertical load. Stress and isostatic patterns

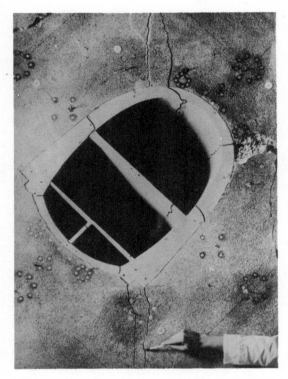

Fig. 140. Cisa Pass tunnel model. Detail of the model after the final test to failure. Load at 45°

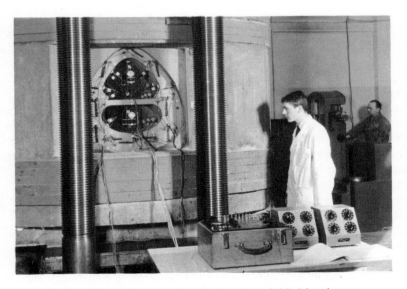

Fig. 141. "Salagou" dam bottom discharge tunnel. Model under test

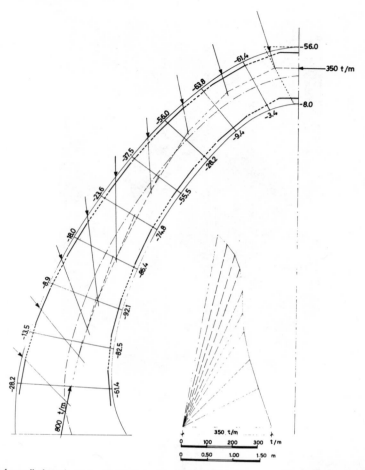

Fig. 142. "Salagou" dam bottom discharge tunnel. Rockfill pressure on liner as derived from the stresses

measured in the lining of the tunnel, is shown in Fig. 142. As can be seen, the loose material in this case was contained in a large reinforced concrete containing structure. To minimise the angle of friction with the wall of the casing the wall was lined with a thin layer of grease contained between two polythene sheets.

Among other laboratories that have been active in the field of geomechanical models is numbered the geotechnical laboratory of the University of Sarajevo under the direction of Krsmanovitch.

The research on plane models of elementary arches of the "Grancarero" dam is noted here as an example of their work.

These models were designed to investigate the stability of the abutments of some arches in relation to the more important of the hypothesis related to considering the dam and its foundation acting as independent arches.

It must be realised that the plane model represents an elementary idealisation of the problem that ignores the more complex and realistic spatial behaviour of

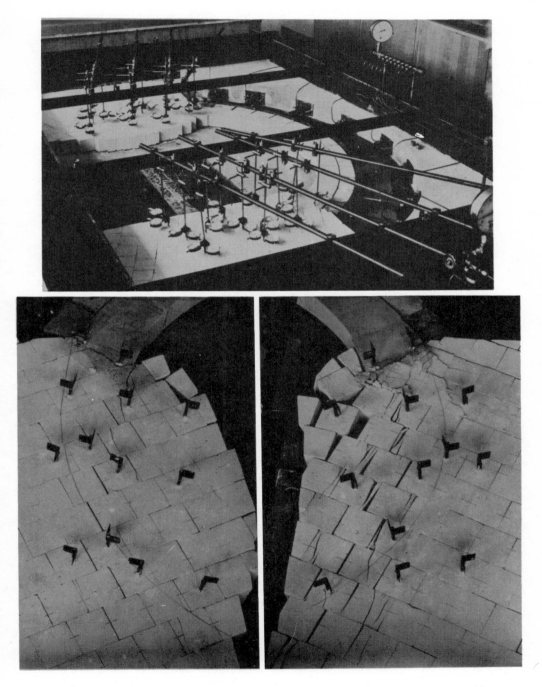

Fig. 143. Two-dimensional geomechanical model of "Grancarevo" dam. Failure due to a combination of sliding and failure of blocks in the abutment

the structure; however it offers the advantage of allowing the examination of the internal deformation distribution and the mode of failure within the rock according to an investigation on horizontal sections.

In qualitative terms these plane models, if executed in parallel with three-dimensional models as in the specific case quoted, can provide complementary information of particular interest, especially in connection with experiments to failure.

The strength capacity of the rock depends on the orientation and eventual imbrication of the elementary blocks and the angle of friction between the surfaces in contact. When the orientation of the blocks does not permit continuous sliding surfaces some blocks may fracture and in this case behaviour is governed by the mechanical strength properties of the material. On this point Fig. 143 shows a failure composed in part of sliding and in part by failure of the blocks [15].

Of particular interest is the path of the surfaces of separation that, to a first approximation, form a failure surface according to the isostatic lines of maximum shear.

Some objections have been made on the static correspondence between plane and three-dimensional models. Without entering into the merits of this argument we think that, taking account of the particular significance normally accorded to the results of these tests, these objections frequently arise from an excessively theoretical rigour and to not take into consideration that in each case the experimental results are more realistic and reliable for the correlation obtained with the traditional analytical procedures.

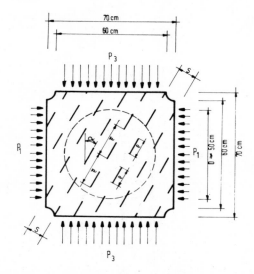

Fig. 144. Two-dimensional model for investigating the behaviour of fissured rock masses. Dimensions of the test blocks. Symbols: e length of outcrop of the minor joints; $\bar{e}$ distance between the joints; p principal load; p_1 transverse load ($|p_1| < |p_3|$); α angle between the direction of the principal load and the joints.

Finally the research carried out by "Interfels" at Salzburg, under the direction of Müller and Pacher, must be noted. These investigations determined the strength of blocks that ideally reproduced, at a reduced scale, variably fissured massive rocks.

The tests were carried out on 70 cm × 70 cm cubes (Fig. 144) [22] varying the conditions of fissuration in the following manner:

angle between the family of fissures and the direction of principal vertical load $\alpha = 0° - 15° - 30° - 45° - 60°$;
ratio of surface apertures $\chi = \rho/\bar{\rho} = 0 - ^1/_3 - ^2/_3 - 1$;
frequency of fissuration $\bar{\rho}/S = 1.2 - 2.4$;
ratio of specific loads $P_3/P_1 = 3 - 5 - 10 - \infty$.

In this way the influence of the fissures on the mechanical strength of the mass was made evident.

In the presence of fissures the strength may be reduced to 15% of that of the monolithic blocks and is very variable from situation to situation. The reductions become greater the nearer the load approaches to a uniaxial state of stress. The aperture ratio causes some reduction in strength for $\chi = ^1/_3$ but its influence increases greatly at values in excess of $\chi = ^2/_3$.

Failure of the block produces for the most part two families of discontinuities: one represented by the existing family of fissures, the other generated during the test to failure and inclined at 70—80° to the first, thus forming a stepped pattern of fissuration.

The overall failure mode suggests a Mohr type failure criterion. Failure is in general of a plastic nature with very high axial deformations. Considerable lateral dilatational deformations were also apparent, as is characteristic of highly fissured media.

These results are in good agreement with in-situ experience, experience gained on a large scale on the rock abutments of "Kurobe" dam (Japan) [31].

Other experiments worth mentioning have been carried out at the "Institut für Bodenmechanik und Felsmechanik" of the University of Karlsruhe, also under the direction of Müller. These concern schematic researches on sliding phenomena of large masses intersected by inclined planes of discontinuity [20].

The researches are intended to examine the kinematics of sliding of discontinuous masses on a surface inclined at a given angle; and ultimately the disturbance generated in the mass on passing from the inclined plane on to a near-horizontal plane. The purpose of this work is to investigate some hypotheses advanced after the sliding failure of Mount Toc (Vajont).

As an illustration of this work Fig. 145 shows the deformation plots of the mass for the test illustrated, from which one can see the relative deformations of original reference circles after sliding.

The relaxation of the rock was measured as the ratio $\delta\, a/a$ (see Fig. 145).

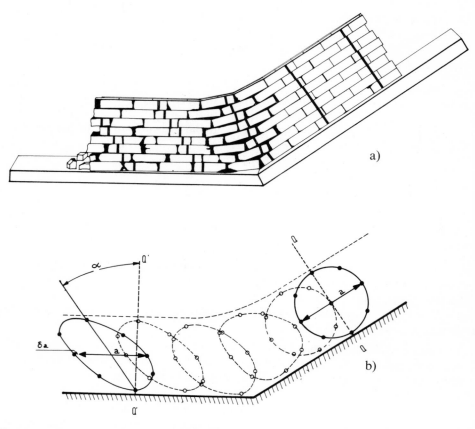

Fig. 145. Sliding experiments: a) relative movements between the layers, opening up of planes of stratification and loosening of the model mass after sliding to a new plane, b) deformation of the model mass during sliding. α = angle of shear deformation due to relative movements of the layers; loosening $L\% = \delta_\alpha/a \cdot 100$

References

[1] Everling, G.: Model Tests Concerning the Interaction of Ground and Roof Support in Gate-roads. Intern. J. Rock Mech. and Min. Sci., Vol. I, 319—326 (1964).

[2] Fumagalli, E.: Model Simulation of Rock Mechanics Problems. Rock Mechanics in Engineering Practice (Stagg, K. G., Zienkiewicz, O. C., eds.). London: J. Wiley. Bulletin Ismes Nr. 38 (October 1968).

[3] Fumagalli, E.: Tecnica e materiali per la modellazione delle rocce di fondazione di sbarramenti idraulici. Bulletin Ismes, Nr. 17 (May 1962).

[4] Fumagalli, E.: Modèles géoméchaniques des réservoirs artificiels: matériaux, technique d'essais, exemples de reproduction sur modèles. Bulletin Ismes, Nr. 26 (October 1964).

[5] Fumagalli, E.: Equilibrio geomeccanico del banco di sottofondazione alla diga del Pertusillo. Bulletin Ismes, Nr. 31 (February 1966).

[6] Fumagalli, E.: Stability of Arch Dam Rock Abutments. Bulletin Ismes, Nr. 32 (May 1967).

[7] Fumagalli, E.: Tecnica de vanguardia en modelacion (modelos geomecanicos). Imme, Nr. 20 (October-December 1967).

[8] Gaziev, E. G., Erlikhman, S. A.: Stresses and Strains in Anisotropic Rock Foundation (Model Studies). Symposium of the International Society for Rock Mechanics, II-1 Nancy (1971).

[9] Goodman, R. E., Taylor, R. L., Brekke, T. L.: A Model for the Mechanics of Jointed Rock. J. Soil Mech. and Found. Div., Proceeding ASCE, Vol. 94, Nr. SM3, 637 to 659 (May 1968).

[10] Jaeger, J. C.: The Frictional Properties of Joints in Rock. Geofis. Pura Appl., Vol. 43, 148—158 (1959).

[11] John, K. W.: Graphical Stability Analysis of Slopes in Jointed Rock. J. Soil Mech. and Found. Div., Proceeding ASCE, Vol. 94, Nr. SM2, 497—526 (March 1968).

[12] John, K. W., Müller, L.: Modellstudien zum Erfassen des geomechanischen Verhaltens von Gebirgsmassen. Veröff. Inst. Bodenmech. u. Felsmech., Techn. Hochsch. Karlsruhe, Heft 24 (1966).

[13] Judd, W. R.: Problems that have Arisen During Ten Years of Rock Mechanics. Festschr. 10 Jahre IGB, 128—151. Berlin: Akademie Verlag. 1968.

[14] Krsmanovic, D., Milic, S.: Model Experiments on Pressure Distribution in Some Cases of a Discontinuum. Felsmech. u. Ing. Geol., Suppl. I, XIV, Wien (1964).

[15] Krsmanovic, D., Tufo, M., Langof, Z.: Shear Strength of Rock Masses and Possibilities of its Reproduction on Models. Proceedings of the 1st Congress of the International Society of Rock Mechanics, Vol. 1, 537—542, Lisboa (1966).

[16] Kuznecov, G. N.: Modellversuche zum Einflusse der Gesteinszerklüftung auf die Pfeilerstandfestigkeit beim Kammerabbau. Bericht 5, Ländertr. Intern. Büro f. Gebirgsmech., Leipzig, 133—149 (1963—1964).

[17] Maury, V., Duffaut, P.: Etude sur modèle des distributions de contrainte en milieux à deux familles de discontinuités. Deuxième Congrès de la Société Internationale de Mécanique des Roches, 8—16, Beograd (1970).

[18] Müller, L.: Untersuchungen über statistische Kluftmessung. Geol. u. Bauw., Jg. 5/4, 185 (1933).

[19] Müller, L.: Der Felsbau, Band I. Stuttgart: Enke. 1963.

[20] Müller, L.: Der progressive Bruch in geklüfteten Medien. Sitzungsber. 1. Kongr. Intern. Ges. Felsmech., Band I, 679—686, Lissabon (1966).

[21] Müller, L., Pacher, F.: Geomechanische Auswertung geologischer Aufnahmen auf der Schachtanlage Neumühl. Unveröff. Ber. (1956).

[22] Müller, L., Pacher, F.: Modellversuche zur Klärung der Bruchgefahr geklüfteter Medien. Rock Mechanics and Engineering Geology, Suppl. II (1965).

[23] Nelson, R. A.: Modelling a Jointed Rock Mass. M. S. Thesis, MIT, Dept. Civil Engrg. (September 1968).

[24] Oberti, G.: Die Theorie der Modelle unter besonderem Hinweis auf Probleme der Geomechanik. Bericht 3, Ländertr. Intern. Büro f. Gebirgsmech., Leipzig, 31—42 (1961—1962).

[25] Oberti, G.: Modelos de presas de concreto y tuneles. Bulletin Ismes, Nr. 37 (December 1967).

[26] Oberti, G., Fumagalli, E.: Results obtained in Geomechanical Models Studies. Bulletin Ismes, Nr. 26 (October 1964).

[27] Oberti, G., Fumagalli, E.: Geomechanical Models for Testing the Statical Behavior of Dams Resting on Highly Deformable Rock Foundation. Rock Mechanics and Engineering Geology, Vol. 1/2 (1963).

[28] Oberti, G., Fumagalli, E.: Propriétés physico-mécaniques des roches d'appui aux grands barrages et leurs influences statiques documentées par les modèles. Huitième Congrès des Grands Barrages, Edimbourg, 1964, R. 35-Q. 28.

[29] Rengers, N., Müller, L.: Kinematische Versuche an geomechanischen Modellen. Rock Mechanics, Suppl. I, 20—31 (1969).

[30] Stimpson, B.: Modelling Materials for Engineering Rock Mechanics. Imperial College, London, Rock Mech. Pres., Report Nr. 8 (July 1968).

[31] Takano, M.: Rupture Studies on Arch Dam Foundation by Means of Models. The Kansai Electric Power (February 1960).

Subject Index

Composed by Austro-Filmsatz Richard Gerin, A-1020 Wien
Printed by Paul Gerin, A-1021 Wien